DNA Tumor Viruses

DNA Tumor Viruses

H. E. Tao
Editor

Nova Biomedical Books
New York

For permission to use material from this book please contact us:
Telephone 631-231-7269; Fax 631-231-8175
Web Site: http://www.novapublishers.com

LIBRARY OF CONGRESS CATALOGING-IN-PUBLICATION DATA

DNA tumor viruses / H.E. Tao (editor).
p. ; cm.
Includes bibliographical references and index.
ISBN 978-1-60692-111-1 (hardcover)
1. Oncogenic DNA viruses. I. Tao, H. E. [DNLM: 1. DNA Tumor Viruses--immunology. 2. DNA Tumor Viruses--pathogenicity. 3. Tumor Virus Infections--genetics. 4. Tumor Virus Infections--immunology. QW 166 D6287 2008]
QR372.O58D53 2008
616.99'4071--dc22

2008034443

Published by Nova Science Publishers, Inc. ✝ New York

CONTENTS

PREFACE

Chapter 1 - Simian Vacuolating virus 40 (SV40) is a small DNA oncogenic virus first isolated in 1960 from contaminated polio vaccines. Soon after its discovery, it was recognized SV40 is highly oncogenic in experimental animals. The types of tumors induced by SV40 depend on the route of viral administration. For this reason, SV40 is classified as polyomavirus. The SV40 genome contains a limited amount of genetic information. Hence, this virus depends almost entirely on the host's biochemical machinery for replication. Nevertheless, SV40 employs specific strategies to deregulate the host's cell cycle to replicate its own genome. SV40 achieves this task through many elegant molecular biology strategies which allow it to maximize usage of the SV40 circular chromosome. More importantly, all protein products coded for the 5-6 SV40 genes are highly multifunctional during each phase of the viral infection. The degree of SV40 proteins' multifunctionality has been extraordinarily exploited in the viral major oncogene (the Large T antigen, or Tag), which interacts with (and influences the activities of) a large number of cellular proteins. Throughout the last four decades, SV40 has been used extensively as a tool for understanding basic molecular biology concepts; yet, this apparently simple virus always reserves new surprises for researchers. In the past decade, either alone, or in cooperation with environmental carcinogens, SV40 has been increasingly linked to human carcinogenicity. Interest in SV40 has increased because of its presence in certain forms of human cancers, including mesothelioma, brain, bone tumors, and more recently, in non-Hodgkin's lymphoma. However, though evidence of SV40's presence has been found in certain tumors, it has not been determined SV40 actually causes these cancers. Population-based studies are conflicting, and more are necessary to address causality. Many epidemiological studies found no increased risk for developing cancer among individuals exposed to SV40-contaminated vaccines; however, all of these studies had significant limitations, including: small sample size of rare tumors, ambiguity in defining exposed and control individuals, and technical difficulties. These limitations make it difficult to draw any definitive conclusions on the increased risk of human cancer associated with SV40-contaminated polio vaccine. Nevertheless, the presence of SV40 DNA sequences in individuals not directly exposed to SV40-contaminated vaccines suggests there may be other modes of viral transmission in human populations which could confound these results. This review presents some basic aspects of SV40's oncogenic potential; its role in cancer; the epidemiological evidence shaping controversies in the field; present efforts; and a general view on the potential future directions.

Chapter 2 - Epstein-Barr virus (EBV), a gamma herpesvirus, is implicated in the pathogenesis of a variety of human malignancies including Burkitt's lymphoma (BL), nasopharyngeal carcinoma (NPC), Hodgkin's disease, and post-transplant B-cell lymphoma. Studies examining the mechanisms by which EBV causes malignancies have focused on those viral encoded proteins that are expressed during latency but there have been few studies to determine the potential role that proteins produced during lytic replication of EBV may have in the pathophysiology of EBV-associated diseases. We have demonstrated that the EBV-encoded deoxyuridine triphosphate nucleotidohydrolase (dUTPase), an enzyme that is produced during lytic replication of EBV, causes immune dysregulation in addition to its enzymatic function. The EBV-encoded dUTPase induces the upregulation of several proinflammatory cytokines including TNF-α, IL-1β, IL-6 and IL-8 as well as IL-10 in human monocytes/macrophages through the activation of NF-κB. Furthermore, the EBV-encoded dUTPase inhibited human T-cell proliferation in vitro. The EBV-encoded dUTPase also inhibited the replication of mitogen-stimulated lymphocytes and the synthesis of interferon-γ by cells isolated from lymph nodes and spleens of mice treated with the dUTPase. The data provide a new perspective on how a nonstructural protein that is associated with lytic replication of EBV can cause immune dysregulation. While the production of these proinflammatory cytokines may contribute to the pathology associated with EBV infection, perhaps more importantly, the production of IL-10 may alter the response of T-cells, ($CD4^+$ and $CD8^+$) to EBV infection. This may create a favorable environment for controlling the steady state expression of latent EBV and the ability of EBV genome positive tumor cells to survive.

Chapter 3 - Epstein Barr virus (EBV) is a DNA virus at can be the cause of several cancers. This DNA virus is proved as DNA tumor virus. Examples of EBV – related cancers include nasopharyngeal carcinoma and lymphoma. The reports on oncogenesis of EBV in significant cancers will be discussed in this chapter. Detection for the EBV is a commonly applied virological laboratory. In this work, the present novel knowledge on EBV laboratory diagnosis, therapy and prevention are reviewed and discussed. In addition, the author also performs a metanalysis on the previous reports on EBV virus infection in tropical countries.

Chapter 4 - Human papillomaviruses (HPV) have been etiologically associated with malignant lesions, most notably cervical, penile, anal and oro-pharynxal. Approximately 40 HPV types have been associated with lesions of the anogenital tract and these can be further classified into so-called low-risk and high-risk types based on their association with clinical lesions. While low-risk HPVs are generally associated with benign lesions such as condyloma acuminatum, infections with high-risk HPVs are associated with a significantly increased risk of developing cancer. Based on epidemiological and moleclular biological evidence, it is widely accepted that high risk HPVs, including HPV types 16 and 18, play an etiological role in the development of cancer. Two oncogenic viral proteins, E6 and E7, involved in cell cycle control, are required for the efficient immortalization of their natural host cells, resulting in primary human squamous epithelial cells developing mucosal and skin cancers. These proteins have been the focus of many studies of cervical cancer during the past 20 years, and their oncogenic mechanisms are now established E7 binds to and degrades the tumor suppressor Rb, while E6 binds to and inactivates the tumor suppressor p53. The causal relationship between chronic inflammation and cancer is widely accepted. Specifically, there is a strong association between tumor viruses and development of cancers. Activation of NF-kB is also involved cancer development and progression. Therefore, this short review focuses on the molecular players during the development from chronic inflammation to cancer.

Chapter 5 - Human papilloma virus (HPV) is a DNA virus that can be the cause of many carcinomas. This DNA virus is proved as DNA tumor virus. The cervical carcinoma is the main carcinoma that relates to HPV infection. Screening for the HPV DNA is the new laboratory analysis in gynecology. Also, the vaccine for HPV is presently available. In this work, the present new knowledge on HPV laboratory diagnosis, therapy and prevention are reviewed and discussed. In addition, the author also performs a metanlysis on the previous reports on cervical carcinoma, the most common female carcinoma in Thailand, and HPV virus infection in Thailand.

Chapter 6 - Kaposi′s sarcoma (KS) associated herpesvirus (KSHV), also termed human herpesvirus 8 (HHV-8) was discovered in 1994 when Chang and colleagues reported the finding of herpesvirus-like sequences in skin lesions from AIDS-KS patients [1]. The virus is present in all KS clinical forms, and it is also associated with another two neoplasic diseases: multicentric Castleman's disease (MCD) and primary effusion lymphoma (PEL). PEL is a rare case of lymphoma first identified as a subset of body-cavity-based lymphomas BCBL [2]. These lymphomas present initially as a lymphomatous effusion located in body cavities without a contiguous tumor mass. PEL cells can be *in vitro* cultured and cell lines PEL-derived are a well-characterized model for a cancer caused by the expression of a few HHV-8 latent proteins [3-5]. Even though most PEL cells are co-infected with HHV-8 and Epstein Baar virus (EBV), some lines, such BC-3, KS-1, BCP-1, CROAP/3, and BCBL-1 contain only HHV-8 episomals [5-9]. Episomes can turn from latent to viral progenie after stimulation with phorbol ester 12- O – tetradenoyl phorbol-13 acetate (TPA), enlarging 16 times the amount of virus [10]. Serum antibodies that are bound to the virus sprouting in the cell are revealed by indirect immunoflourescence assay (IFA). The use of corticosteroids in transplanted patients increases the occurrence and severity of KS [11]. *In vitro,* hydrocortisone induces a viral lytic reactivation in BCBL-1 cell line [12]. In our laboratory, we have examined the ability of dexamethasone (DEX) to induce HHV-8 lytic phase on BCBL-1. We compared the action of DEX with TPA on cell viability, amount of antigen expression, and the performance of IFA slides to detect antibodies against HHV-8 in KS patients, HIV infected individuals and blood donors. Results were analysed by Fisher exact test. Viable cells were counted daily by trypan blue exclusion for six days. Differences between TPA-induced and control cells were found, but no differences were seen in DEX vs. control cells. Daily viral antigen expression within the cells was examined by IFA using HHV-8 positive pool antisera [13]. Slides containing TPA, or DEX-induced cells were made up and cells expressing the HHV-8 lytic pattern were counted by IFA [14]. No differences ($p=0.67$) were observed among the amount of fluorescent cells in slides with TPA-or DEX-induced cells. Also, no differences were observed between TPA and DEX on fluorescent pattern. In order to investigate if DEX induced cells may be used for HHV-8 serological diagnosis, 86 HHV-8 positive sera from previously described populations were IFA tested and titered in TPA and DEX slides; 54 HHV-8 negative sera were also included [14]. Results fully matched both slides. Our data demonstrate that DEX could be used for studies of HHV-8 lytic cycle activation events and as an alternative to TPA induction to make HHV-8 IFA slides for serologic diagnosis.

Chapter 7 - The polyomavirus family includes two members, BK virus (BKV) and JC virus (JCV), that naturally infect humans. Besides these, the primate polyomavirus, simian vacuolating virus (SV40), was isolated from humans as a consequence of contaminated polio vaccines in the 1950's in theUnited States. Human polyomaviruses are ubiquitous in the human population worldwide and cause disease only in immunocompromised individuals: progressive multifocal leukoencephalopathy (PML) mainly in AIDS patients and polyomavirus-associated nephropathy (PVAN) in renal transplant patients, being the etiologic agent JCV and BKV, respectively. Many studies indicate that these three polyomaviruses are associated with cancers in humans as their genome sequences and protein expression are detected in several types of human neoplasm. Polyomaviruses rely on cellular enzymes and cofactors for DNA replication and these proteins are expressed in S phase; then JCV, BKV and SV40 encode regulatory proteins (large T antigen, middle T antigen and small t antigen) that drive resting cells into S phase through interactions with cell cycle regulators such as pRb and p53. This aberrant stimulation of the cell cycle is a driving force of oncogenic transformation. Here we describe what is known until now about the molecular mechanisms of these interactions.

Chapter 8 - Epstein-Barr virus (EBV) is known as a causative agent of Burkitt's lymphoma, nasopharyngeal carcinoma, and about 10% of the cases of stomach carcinoma. We have reported EBV gene expression including EBV lytic infection protein in many other human cancers. We have also identified the expression and replication of EBV genes in normal human macrophages, and abnormal histiocytes in Langerhans' cell histiocytosis (LCH). Furthermore, we revealed EBV expression in macrophages in the lesions of LCH, primary lung lymphoma, uterine cervical carcinoma, renal cell carcinoma, tongue cancer, cutaneous T-cell lymphoma, and thyroid carcinoma. Therefore, tumor-associated macrophages in EBV-related tumors are likely to carry EBV, and they appear to induce EBV lytic infection. Lytic infection of EBV in multiple tissues may provoke a strong inflammatory response, because cell lysis induced by virus replication results in intense immune responses against viral proteins. We reported that childhood LCH expressed high level of EBV lytic infection protein, and that, in one case, the administration of acyclovir resulted in complete remission. Anti-herpesvirus drugs seem to be effective against the disease. We reported that uterine cervical carcinoma expresses EBV genes which participate in transformation, and that carcinoma cells of almost all cases of cervical carcinoma also carried the HPV 16 gene. It is interesting that EBV and HPV may cooperate in the development of cervical carcinoma through the long-term inflammation caused by chronic infection with both viruses. The roles of macrophages not only in inflammation but also in the interaction between viruses or viruses and microbes should be clarified. Through such studies of inflammation involving macrophages, the dynamic mechanism leading to EBV oncogenesis will be elucidated.

In: DNA Tumor Viruses
Editor: H. E. Tao
ISBN 978-1-60692-111-1

Chapter I

SV40 AND CANCER

Maria E Ramos-Nino[1] and Maurizio Bocchetta[2]
[1]Department of Pathology, University of Vermont College of Medicine, Burlington, Vermont 05405, USA;
[2]Thoracic Oncology Program, Cardinal Bernardin Cancer Center, Loyola University, Maywood, Illinois 60153, USA.

ABSTRACT

Simian Vacuolating virus 40 (SV40) is a small DNA oncogenic virus first isolated in 1960 from contaminated polio vaccines. Soon after its discovery, it was recognized SV40 is highly oncogenic in experimental animals. The types of tumors induced by SV40 depend on the route of viral administration. For this reason, SV40 is classified as polyomavirus. The SV40 genome contains a limited amount of genetic information. Hence, this virus depends almost entirely on the host's biochemical machinery for replication. Nevertheless, SV40 employs specific strategies to deregulate the host's cell cycle to replicate its own genome. SV40 achieves this task through many elegant molecular biology strategies which allow it to maximize usage of the SV40 circular chromosome. More importantly, all protein products coded for the 5-6 SV40 genes are highly multifunctional during each phase of the viral infection. The degree of SV40 proteins' multifunctionality has been extraordinarily exploited in the viral major oncogene (the Large T antigen, or Tag), which interacts with (and influences the activities of) a large number of cellular proteins. Throughout the last four decades, SV40 has been used extensively as a tool for understanding basic molecular biology concepts; yet, this apparently simple virus always reserves new surprises for researchers. In the past decade, either alone, or in cooperation with environmental carcinogens, SV40 has been increasingly linked to human carcinogenicity. Interest in SV40 has increased because of its presence in certain forms of human cancers, including mesothelioma, brain, bone tumors, and more recently, in non-Hodgkin's lymphoma. However, though evidence of SV40's presence has been found in certain tumors, it has not been determined SV40 actually causes these cancers. Population-based studies are conflicting, and more are necessary to address causality. Many epidemiological studies found no increased risk for developing cancer among individuals exposed to SV40-contaminated vaccines; however, all of these studies had significant limitations, including: small sample size of rare tumors, ambiguity in defining exposed and control individuals, and technical difficulties. These limitations make it difficult to draw any definitive conclusions on the increased risk of human cancer associated with SV40-contaminated polio vaccine. Nevertheless, the presence of SV40 DNA sequences in individuals not directly exposed to SV40-contaminated vaccines suggests there may be other modes of viral transmission in human populations which could confound these results. This review presents some basic aspects of SV40's oncogenic potential; its role in cancer; the epidemiological evidence shaping controversies in the field; present efforts; and a general view on the potential future directions.

INTRODUCTION

Simian vacuolating virus 40 (SV40) is a double-stranded DNA polyomavirus discovered more than 40 years ago. SV40 is very similar in size, genome organization, and DNA sequence [1] to other polyomaviruses found in humans, such as JC virus and BK virus. Unlike them, SV40's natural host is the rhesus monkey, not the human. The first reports of SV40's association 40 with human cancer were published in the early 1970s [2-4]. Evidence of this association has been mounting (reviewed in [5-9]). The potential involvement of SV40 in human cancers is crucial due to the inadvertent exposure of nearly 100 million people in the United States [7] and many more in other countries. SV40 contaminated both the inactivated poliovirus vaccine (IPV or Salk vaccine) and the live attenuated oral poliovirus vaccine (OPV or Sabin vaccine) produced in primary kidney cells from rhesus and cynomolgus macaques [5]. In addition, different adenovirus vaccines used on a limited scale for United States military personnel from 1961 to 1965 also contained live SV40 [10]. Vaccination with IPV from 1955 to 1963 is thought to be the primary source of human exposure to SV40 in United States [7]. The large-scale exposure of the human population to the virus, together with the extensive evidence of its oncogenic potential in *in vitro* and animal models has raised concern regarding the public health risk of SV40 exposure.

MECHANISMS OF VIRAL INFECTION

Summary of the Viral Structure

Simian vacuolating virus 40 (SV40) is a non-enveloped DNA virus first isolated from contaminated poliovaccines [11]. The viral particles have a sedimentation coefficient of 240 S in sucrose gradients [12,13].The capsid is composed of 72 pentamers (each pentamer forms a capsomer) of the main viral structural protein VP-1, while the center of the pentamer is occupied by either VP-2 or VP-3 proteins [14]. VP-1 appears to be responsible for the overall structure of the viral envelope, as VP-1 synthesized either in *E.coli* or in insect cells assembles into typical viral-like particles very similar to genuine SV40 virions in a calcium-dependent fashion [13,15]. The individual capsomers are held together by the carboxy terminal sequence of VP-1. Some authors have proposed these viral-like particles as promising vectors for gene transfer delivery, since they can potentially accommodate a sizable amount of genetic material [16-18]. In vivo, the icosahedral capsid envelops the double-stranded SV40 circular genome, which is about 5.2 Kbp long ([19]; the exact length of the SV40 chromosome varies among different strains of SV40; [10]). The viral DNA is wrapped into nucleosomes, with all nucleosomal histones being packaged alongside the viral DNA [20,21].

Viral Pattern of Infection and Cell Entry

SV40 infects cells from different species over a broad spectrum of tissues. The outcome of the infection is species-dependent. Rodent cells are termed non-permissive to SV40 infection because SV40 is unable to replicate its genome in these cells. The latter phenomenon is due to the inability of the SV40 replicase (the SV40 Large T antigen, or Tag, see below) to interact with the host DNA polymerase I-α primase [22-24]. Therefore, SV40 enters rodent cells, and successfully translocates its genome into the host nucleus, where its early genes are transcribed and synthesized. However, this process gives rise to an abortive infection with no (or very little) viral replication, and lack of progeny viral particles. On the other hand, monkey kidney cells are fully permissive to SV40 infection in vitro. After viral entry and synthesis of the SV40, early gene products the host cells forcedly enter the cell cycle, with the consequent replication of the SV40 chromosome. This event is followed by a large accumulation of SV40 viral particles and cell lysis (this chain of events will be described in more detail below). Productive SV40 infection does not necessarily require death of the host cell, since some SV40 infected epithelial cells can release large amounts of SV40 particles in the absence of cell lysis [25]. However, necrotic cell lysis is the most common outcome of permissive SV40 infection.

Human cells are generally semi-permissive; indicating only a fraction of a human cell population exposed to infectious SV40 at any given time will support early protein expression followed by SV40 DNA replication and viral particle production, with consequent cell lysis and release in the extracellular environment of progeny SV40 virions. The rest of the SV40-exposed cell population usually displays no sign of viral infection. This occurrence appears to depend on cellular factors which possibly facilitate the correct translocation of the SV40 genome to the nucleus, or the expression of the SV40 early proteins, since all human cells exposed to SV40 virions bind and internalize SV40 particles [26]. There are few departures from this paradigmatic infection pattern of human cells, which are best exemplified by human primary mesothelial cells. The latter are extremely susceptible to SV40 infection and SV40-mediated transformation; nonetheless, infected mesothelial cells appear resistant to SV40-induced cell lysis through a mechanism apparently involving the host's p53 [27]. Intriguingly, SV40 has been most convincingly associated with mesothelioma, a cancer arising from the malignant transformation of mesothelial cells.

The cellular receptor for SV40 particles is the ubiquitous Major Histocompatibility Complex of Class I (MHC-I) [28,29]. However, there are several sources of experimental evidence suggesting HMC-I is necessary, but insufficient for SV40's infection of cells. For example, polarized Vero C1008 monkey kidney epithelial cells bind SV40 at the apical surface, and not at the basolateral membrane, even though the expression of MHC-I in these cells is approximately uniform around the plasma membrane [30].

After binding to its receptor(s) SV40 is internalized into uncoated vesicles which eventually fuse with the endoplasmic reticulum (ER) [31]. This absorption process was later assigned to the caveolae-based route of cellular endocytosis [32]. Indeed, HMC-I molecules are mostly localized in caveolar/lipid rafts [33,34], and caveolae are involved, among other phenomena, in the process of reverse endocytosis, which determines the trafficking of plasma membranes back to the ER. SV40 undergoes at least a partial uncoating in the ER [34,35]. It should be noted caveolin-independent SV40 phagocytosis has been described in human hepatoma 7 cells [36]. Even in this case, the internalized SV40 particles are eventually trafficked to the ER. After being processed in the ER, SV40 DNA translocates to the nucleus through still unclear mechanisms. Some evidence suggests the SV40 genome finds its way into the nucleus through the nuclear pores [37]; however it is still uncertain whether SV40 reaches the nuclear pores as completely uncoated DNA, or whether capsid components (such as VP-3) facilitate nuclear entry and cross the nuclear membrane alongside the SV40 chromosome [38]. Some studies have underscored the importance of VP-1 and VP-3 for viral association with importins and nuclear receptors [39]. Moreover, isolated nuclei required wild-type VP-3 for internalization of SV40 particles [40]. Therefore, it is likely SV40 virions undergo a partial uncoating in the ER, but that incomplete SV40 particles with some capsid components are necessary for SV40 entry in the nucleus, where the SV40 chromosome is eventually freed of most capsid proteins. It should be noted completely "naked" SV40 DNA is not necessarily required for its efficient transcription. Minor SV40 capsid proteins VP-2 and VP-3 have been shown to bind to the SV40 regulatory regions, while interacting with transcriptional factors [41]. Thus, it is possible these capsid proteins may assist cellular factors in regulating the SV40 DNA transcription [42]. Once in the nucleus, the SV40 DNA is transcribed to initiate the early phase of infection.

SV40 Genomic Organization and early Proteins

The SV40 minichromosome comprises a regulatory region, including a bidirectional promoter and the viral origin of replication, and two gene encoding regions (the "early" and "late" regions) spanning in opposite orientations with respect to the regulatory region. At the 3' end of these gene-harboring DNA sequences, there are two closely spaced polyadenylation signals (terminating the early and late transcripts, respectively). The naming of the latter DNA regions reflects the order of transcription after SV40 infection. Since transcription of the late region requires efficient SV40 DNA replication [43], non-permissive rodent cells do not sustain an SV40 late phase infection. Transcription of the SV40 early and late regions is tightly regulated and chronologically separated during SV40 lytic infection. This avoids production of potential antisense RNA which may trigger either interferon-like response, with consequence activation of protein kinase R (PKR), leading to general translational repression and induction of apoptosis [44], or dicer-mediated RNA interference.

The SV40 genome contains limited amounts of genetic information. Therefore, the virus relies heavily on the host's proteins for transcription and replication. However, SV40 has developed a number of strategies to expand the amount of genetic information contained in its chromosome by exploiting phenomena, including alternative splicing of primary transcripts and partially overlapping alternative reading frames during the translation process [45]. The early genes include Tag and the small t antigen (or tag), which mRNAs originate from alternative splicing of the same primary transcripts [46,47]. In certain cells, the primary early transcripts are further alternatively spliced to yield additional proteins, such as super T in rodent cells [48], 17K and small leader protein [49,50]. However, these SV40 proteins have not been detected in all cell systems, and their role remains unclear or controversial. Tag, the major SV40 oncoprotein, is an extraordinarily multifunctional protein. It regulates SV40 transcription [51] and initiates viral DNA replication [52]. Tag also helps viral DNA synthesis by allowing the SV40 DNA replication fork to proceed due to the Tag ATP-dependent DNA helicase activity [53]. Tag also influences and interacts with the functions of many of the host's proteins, including: DNA Polymerase α [54], TATA binding protein (TBP) and associated factors (TAFs) [55], the transcriptional co-activator, histone acetyltransferase p300/CBP [56], pRb protein family members [57], heat shock protein 70 (HSP70) [58] and HSP90 [59], the tumor suppressor genes cullin E3 family members and SCF(Fbw7) ubiquitin ligases [60,61], Nbs1, a protein involved in DNA replication control [62], certain members of the p53 protein family [63,64] and other cellular proteins. Tag, either alone or in cooperation with tag, promotes the transactivation of a number of cellular anti-differentiation and/or proliferative genes, including Notch-1 [65], insulin-like growth factor 1 (IGF-1) and its receptor [66,67], the cyclin-dependent kinase cdc2 [68] and c*met*, or hepatocyte growth factor receptor [69]. Tag's coordinated activities are aimed at forcing the cell to enter the cell cycle. This is necessary to assure SV40 chromosome replication, since SV40 relies on cellular proteins and substrates to synthesize its components.

Tag's activities are complemented by tag. The SV40 tag's best-known activity is to bind and inactivate the cellular tumor suppressor gene product protein phosphatase 2A (PP2A) [70,71]. By doing so, tag reinforces mitogenic stimuli through various mechanisms, reflecting the broad array of targets of PP2A. Examples are: the tag induction of the G1/S cyclin D1 and S phase cyclin A1 through the ERK kinase cascade, reinforced by tag-mediated PP2A inhibition [72,73], NFκB induction [74], indirect hyperphosporylation of pRb [75], the Pim-1 protein kinase activation, which exerts its activities alongside myc-regulated transcription [76], the prevented dephosphorylation of cyclin-dependent kinases [77], etc. The SV40 tag also appears to provide essential pro-survival signals, preventing programmed cell death. Although it is still debated whether Tag possesses pro- or anti-apoptotic functions, it is now sufficiently demonstrated tag protects from programmed cell death through the Akt-mTOR signaling pathway [78,79], and possibly, by depressing the activity of the antiapoptotic Bax protein [80].

Dynamics of Infection

As previously mentioned, transcription of the early and late regions are temporally separated in permissive cells. This phenomenon is mediated by a twofold mechanism during the early phase of SV40 infection: Tag-mediated enhancement of early region transcription [81], and, more predominantly, because the late promoter is repressed at two sites, called the +1 and +55 hormone responsive elements (HRE). The numbering reflects the positions on the late primary transcripts. The main repression appears to be operated at the +1 HRE, bound by a number of nuclear receptors, with heterodimers of retinoic X receptor α (RXR-α) and thyroid receptor α (T_3R-α) representing the main transcriptional repressors [82,83]. The +55 HRE also plays an important role in repressing late gene transcription during the early phase of viral infection. It is bound by several hormone receptors, including: the TR orphan receptor family members, COUP-TF receptors, and steroid hormone receptors [84,85].

Most of the SV40 chromosome replication takes place during a prolonged G2 phase, when SV40 genomes account for about 30% of the total cellular DNA. Because the host's DNA content at this point is more than tetraploid, this is often referred as the >G2 phase [86]. Indeed, permissive, SV40-infected cells are G2 arrested during the late phase due to hyper-activation of cellular checkpoint kinase 1 (Chk-1) [87]. During the >G2 phase, SV40 genomes rapidly outnumber the cellular repressors binding the late promoter. This titration effect determines the early-to late switch in SV40 infection [43] with the transcription of the SV40 late genes taking place. This leads to massive accumulation of SV40 late proteins, including (besides VP-1 to –3) agnoprotein, a small, ancillary protein. A micro RNA (miRNA) precursor, encoded in the late transcripts, which targets the early transcripts further removes residual early mRNAs, is eliminated by destruction along the Drosha-Dicer-RISC-mediated RNA interference pathway [88].

Virions are assembled through a mechanism initiated by VP-2 and -3 which binds to six tandem GC boxes (SP-1 recognition elements) in the SV40 regulatory region, and subsequently recruits VP-1 pentamers on this region. Since the host's genome does not contain 6 GC boxes repeated in tandem, only SV40 DNA is encapsidated [41]. The final step of SV40 infection involves activation of cellular poly (ADP-ribose) polymerase (PARP) by VP-3, with consequent host necrosis and release of infectious SV40 particles in the extracellular environment [89,90]. From this infection pattern, it seems intuitive SV40 does not require cell transformation to initiate its life cycle. Probably, the latter may be selectively disadvantageous for SV40, since during mitosis there is no viral replication. SV40-mediated cell transformation may be just a relatively infrequent "accident" during infection deriving from the SV40 oncogenes's activities.

SV40 Oncogenes and their Interaction with Host's Tumor Suppressor Genes

As mentioned, SV40 expands its limited genetic information by using plasticity during the splicing of its transcripts, while using alternative reading frames of its mRNAs during protein translation. Such expansion of viral functions is further exploited in its multifunctional, encoded proteins. Agnoprotein has been involved in facilitating nuclear segregation of SV40 capsid proteins [91], but has also been involved in facilitating viral replication and oncogenesis [92,93]. The JCV (a DNA tumor virus highly homologous to SV40) agnoproteins has been proposed to interact with cellular p53, and to play a role in the pathogenesis of medulloblastomas [94]. VP-2 and VP-3 are also multifunctional, because VP-3 overstimulates PARP activity; leading to viral induced host's necrosis, facilitating viral release in the extracellular environment [89]. VP-2 and VP-3 interact with importins, mediating viral DNA entry into the nucleus [39], and interacting with SP-1 transcription factor to promote viral transcription and packaging [41,42]. Nevertheless, multifunctionality is exploited the most by the SV40 Tag. Some of Tag's functions have been described previously in this review. Here, we will focus on the interactions Tag establishes with the two major tumor suppressor gene products of the cell: p53 and pRb.

Tag binds the unphosphorylated form of the retinoblastoma cellular tumor suppressor gene product [57,95] through a LXCXE motif (residues 103-107). The latter binds the so-called "pocket" domain of pRb protein family members (p107, p110 and p130, or RbB), a domain targeted by several DNA tumor viruses's oncoproteins, including the adenovirus E1A and the papilloma virus E7 proteins. Within the generally accepted model, the binding of Tag to pRb members mimic the cyclin-dependent kinase mediated phosphorylation of pRb proteins. After binding to Tag, retinoblastoma proteins dissociate from E2F-DP complexes, which in turn promotes the transcription of a number of genes involved in G1 progression, including cyclin E [96]. However, such a model is most likely an oversimplification, since retinoblastoma proteins cannot be described as just inhibitors of E2F-DP complexes which allow the latter to function only upon pRb exit from the complex. The "inhibitory" pRb/E2F-DP complex itself also includes SP-1 and HDAC members [97]. Furthermore, retinoblastoma proteins can interact with E2F-DP complexes through domains other than the LXCXE-containing pocket domain [98] and have been described to interact with a number of other cellular proteins, such as Jumonji domain 2 (JMJD2) family of histone acetyl transferase proteins [99], p300/CBP proteins [100], and possibly, 14-3-3 members through DP proteins [101]. Therefore, it is likely Tag's association to pRb members alters the mutiprotein association of retinoblastoma, E2F-DP and other binding partners, by: 1) changing the composition of the complex (e.g., co-repressors leaving the complex, while co-activators entering it), or 2) promoting some structural rearrangements within the complex. The net result promotes a change in pRb/E2F-DP activity from a transcriptional repressor to a transcriptional activator.

The biological significance of Tag's interaction with p53 appears even more complex. In the generally accepted model, Tag, through its binding to cellular p53, inactivates the latter's ability to function as a DNA-dependent transcription factor. Hence, Tag/p53 complexes are generally viewed as "inactive". This interpretation stemmed from studies conducted during the first half of the Nineties: Tag suppressed p53 binding to human ribosomal gene cluster and mouse muscle creatine kinase sequences when baculovirus-synthesized proteins were assayed by EMSA, filter binding and DNAse I protection assays [102]. Similarly, baculovirus-synthesized Tag inhibited binding of a p53-GAL4 fusion protein (overexpressed in *E.coli*) to GAL-4 binding sites [103]. Analogous results were obtained in transfection experiments performed in p53 deficient human osteosarcoma Saos-2 cells. Co-expression of

Tag together with p53 abrogated p53 binding to a 44 bp p53 binding site [104] (Saos-2 are also pRb null, an important issue that will appear below). Collectively, these data strongly suggested Tag inhibits cellular p53 by preventing it from binding to DNA, thus rendering p53 inert, providing the functional equivalent of a p53 knock-out. However, a number of studies indicated the implications of Tag and p53 interactions cannot be fully explained using this model. First, in SV40-transformed cells of different origin, there is evidence of metabolically stable p53 unbound to Tag [105,106], the functions of which still need to be elucidated. More recent studies have shown that p53 complexed to Tag can still bind its cis binding sites specifically [107-109]. Furthermore, Tag-bound p53 extracted from Tag-expressing monkey and human cells was shown to still stimulate transcription of a p53-regulated promoter in cell free extracts [108]. All these studies indicated differences in properties of Tag/p53 complexes obtained from different species, reconstituted Tag/p53 complexes used in vitro translated protein or proteins expressed in *E.coli*. Additionally, a number of studies have shown unexpected cooperation between p53 and Tag in promoting cell transformation. Transformation of rat embryo fibroblasts by SV40 was enhanced by wild type mouse p53 [110]. Similarly, transformation of rat fibroblasts required both Tag and a metabolically stabilized p53 [111]. Transgenic mice expressing Tag in β cells of their pancreatic islets (these mice are susceptible to insulinomas) developed smaller tumors when crossed with p53 deficient mice [112]. Reduced tumor size in Tag expressing, p53 null mice was due to the decreased proliferation of tumor cells, while the apoptosis rate in p53 expressing insulinoma cells was similar to that of p53-deficient tumor cells [112]. These results suggested Tag can be more oncogenic in the presence, rather than in the absence of, wild type p53. These data also suggested Tag's binding to cellular p53 could have conferred some sort of "gain of function" to the latter, which favors cellular transformation, at least in certain cell types. This was confirmed recently in human primary mesothelial cells (HM) and primary astrocytes. Both human cell types form foci of transformed cells 2-6 weeks after SV40 infection [27,113]. These foci can be cultured into immortal cell lines displaying a fully transformed phenotype in vitro driven by the SV40 oncogenes only (e.g., these cells are not exposed in parallel to chemical carcinogens and are not transfected with active oncogenes or telomerase). In these cells, Tag/p53 complexes directly regulate the transcription of the insulin-like growth factor 1 (IGF-1) gene through binding the IGF-1 promoter in a macromolecular complex which includes pRb and p300 [66]. Upon p53 depletion, this complex undergoes a number of modifications, including p300 exit and structural rearrangements of the complex that ultimately lead to transcriptional repression of the IGF-1 gene. This process, in turn, regulates the expression of the IGF-1 receptor (IGF-1R), since it appears IGF-1R is greatly influenced by the IGF-1 autocrine feedback loop in SV40-transformed HM [66]. The IGF-1 signaling pathway is critical for SV40-mediated transformation, because SV40-induced cell proliferation requires the interaction of IGF-1 with its receptor [114]. Therefore, it appears Tag/p53 complexes, through their induction of the IGF-1 gene transcription, play an active role in favoring cell cycle progression of SV40 infected cells (see Figure 1 for a summary of the SV40 Tag interactions with pRb and p53 and the resulting biological effects). This latter mechanism would explain why a Tag N-terminal fragment is still able to bind pRb which failed to produce a morphological transformation when transfected into mouse and rat immortalized embryo fibroblasts alongside a dominant negative p53 [115]. In conclusion, increasingly convincing evidence shows: 1) Tag/p53 complexes actively participate in promoting cell cycle progression (or cell transformation) of the host, and 2) Tag's binding to p53 does not merely inactivate p53 tumor suppressor functions but rather, provides functional gains, is increasingly convincing.

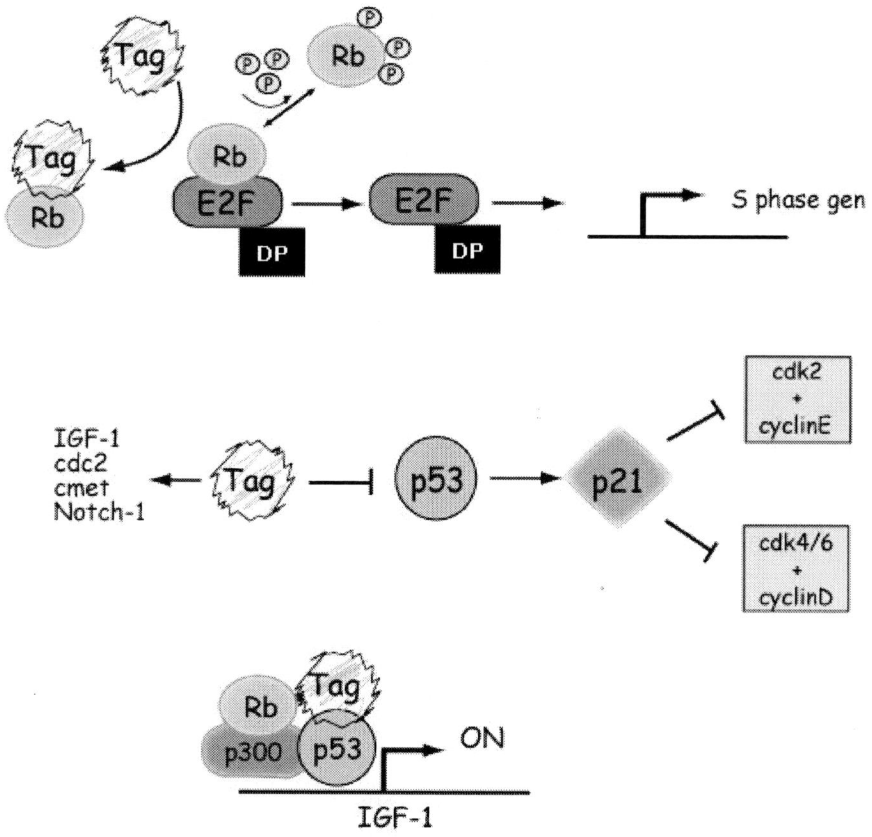

Figure 1. Biological consequences of Tag interactions with the two major cellular tumor suppressor gene products. Tag binding to pRb protein family members causes the activation of E2F/DP transcription factors that in turn promote transcription of genes that cause cell cycle progression (e.g., cyclins, c-myc, etc.). This activity is reminiscent of cyclin-dependent protein kinases (CDK) phosphorylation of pRb family members. Tag interaction with p53 abolishes p53 ability to induce the CDK inhibitor $p21^{CIP1/WAF1}$. Therefore, this event further pushes the cell to go beyond the G1 restriction point. Tag/p53 complexes bind the IGF-1 promoter and stimulate its transcription. Tag also transactivates a number of genes that promote cell cycle progression.

SV40-Mediated Cell Transformation

Because of the mechanisms of infection, SV40 does not transform monkey kidney cells (fully permissive to SV40 infection). Infection of these cells with live SV40 virions always results in the complete destruction of the entire infected monkey kidney cell population. In vivo, SV40 could have potentially pathogenic effects, since the virus establishes latency (through unknown mechanisms) in certain monkey tissues. However, SV40 appears to be pathogenic only in immunocompromized monkeys [116]. This appears to be generally true for polyomaviruses (such as SV40, JCV and BKV). All of these highly homologous DNA oncogenic viruses do not cause major disease in the natural hosts, unless the latter is immunocompromized (e.g., progressive multifocal leukoencephalopathy -PML- caused by JCV in AIDS patients [117]. Another feature of polyomaviruses is their high oncogenicity when inoculated into species other than their natural hosts.

Introduction of the SV40 oncogenes in rodent cells, obtained either through infection or transfection, causes cell transformation at low frequencies. This phenomenon takes place when the viral chromosome, or a plasmid expressing Tag and tag, randomly integrate in the host genome so the early genes can be expressed. This condition recapitulates SV40-mediated transformation of human cells obtained with replication-defective genomes [118]. The SV40 oncogenes, or the human papilloma virus 16 E6 and E7 oncoproteins, have been used as the most common tools to establish transformed/immortal cell lines from virtually any rodent and human tissue. To avoid the collateral cell death often occurring when infecting human cells with live SV40, human cell lines have most often been derived after transfection of replication defective SV40 genomes. Nonetheless, human cell transformation can also be attained by direct infection with live SV40. It must be underscored that efficient human cell transformation requires the activities of both Tag and tag [119].

Primary human cells have a limited life span in culture [120]. After a number of cell divisions (the precise number varies in cells from different tissue origin), the cell population undergoes senescence, characterized by a massive G1 arrest originating from the accumulation of the cyclin dependent kinase inhibitor (INK4A1) p16 and the tumor suppressor gene ARF, also originating from the INK4A locus [121]. In human fibroblasts, which replicate in vitro for about 40-50 cell doublings, expression of the SV40 oncogenes leads to the so-called "proliferative expansion", which increases the fibroblasts replicative potential to about 60-70 cell doublings [118]. Despite this prolonged growth potential, SV40-containing human (and mouse) fibroblasts are nevertheless poised to enter the so-called crisis, an event ultimately triggered by telomeres shortening beyond a critical length [122]. Cells undergoing crisis are either terminally growth arrested (displaying a typically enlarged, irregular morphology with giant cytoplasm expanded in irregular interdigitations), or die of programmed cell death [123]. A small fraction of cells that have entered crisis (on average $1/10^7$;[118]) escapes this condition to become an immortal cell line. All human cells, with one notable exception, follow this paradigm, regardless of whether they display a completely transformed phenotype in vitro. We underscore that cells derived from different tissues have different susceptibility to SV40-mediated transformation after infection. The evidence thus far strongly suggests the rate of SV40 replication in infected cells plays a central role in determining the cell susceptibility to SV40-mediated transformation [27,113,124].

Historically, human cells have been transformed mainly through transfection with replication-defective SV40 genomes. In principal, this system is identical to the model of SV40-mediated transformation of rodent cells; requiring the integration of the SV40-bearing DNA into the host genome while preserving the expression of the SV40 oncogenes. Whether SV40 integrates randomly, or by preferentially targeting certain chromosomes has been the object of some past debate. Current belief is that SV40 can integrate randomly in the human genome. However, several studies suggest chromosome 7 may be a preferred site of SV40 integration [125-128]; and possibly chromosome 17 [129]. Regardless of the insertion chromosome, SV40 integrates preferentially in the proximity of nuclear matrix attachment regions [130]. Occasionally, human fibroblasts can contain both integrated and episomal SV40 DNA that does not recircularize from integrated SV40 genomes due to recombination [131], and some have described SV40-immortalized human fibroblasts as continuously releasing infectious SV40 [132]. In conclusion, there is plenty of evidence human cells can sustain the presence of non-integrated, episomal SV40 DNA while, at the same time, avoiding a fully productive SV40 infection cycle, which would ultimately promote necrosis of the host. The presence and stability of episomal SV40 in human cell cultures and human tumors has been the object of interesting debates in the last decade or so. SV40 has been only sporadically detected as integrated in human malignancies [133]. On the other hand, there is evidence of episomal SV40 in human malignancies (in many studies SV40 was detected in human cancers only when DNA extraction techniques that recovered low molecular weight DNA were employed); and episomal SV40 has been detected in brain tumors [134].

Primary human mesothelial cells (HM) are an exception to the usual pattern of SV40 infection and transformation of human cells. Albeit HM are exceedingly susceptible to SV40 infection, as they support low level viral replication if compared to primary fibroblasts, for example. This phenomenon appears linked to unusually high p53 expression levels in HM [27]. Such reduced SV40 production allows HM to survive SV40 infection, while exposing them to a prolonged period of the SV40 oncogenes's transforming activities. This leads to a uniquely high rate of cell transformation following SV40 infection of HM. About 1/5,000 of infected HM will develop into a tri-dimensional focus of transformed cells [27,113]. Transformed foci can be easily (with more than 95% success rate) cultured into immortal, anchorage-independent cell lines. In tissue culture, these cells release low amounts of infectious SV40. These cell lines are immortal because SV40 infection of HM causes telomerase reactivation early after infection, and the telomerase activity increases with cell passage [135]. Furthermore, the level of telomerase activity appears to correlate with Tag expression levels, since transfection of non-replicating plasmids expressing Tag and tag cannot induce telomerase activity [135]. The latter phenomenon easily explains why SV40-transformed HM derived from transfection of replication-defective SV40 genomes undergo crisis similar to that observed in fibroblasts [27]. From these and other observations, it appears a moderate level of intra-host viral replication is key for SV40-susceptibility to transformation and the generation of immortal clones at high frequency. This concept was confirmed when HM were infected either with SV40 or BKV. The latter is a human polyomavirus which shares an identical genome organization with SV40, a high degree of nucleotide identity, and a highly homologous Tag. BKV infection of HM elicits very similar molecular changes otherwise observed after SV40 infection of HM, including *met* phosphorylation, telomerase induction and Notch-1 activation [113]. However, BKV does not transform infected HM, because it replicates much faster than SV40. This characteristic invariably leads to HM cell death [113]. In conclusion, it appears the host cell's rate of viral replication plays a central role in determining viral-mediated cell transformation after infection. Without viral replication, Tag expression levels are insufficient to induce telomerase, and thus, immortalize human cells. This is not such a crucial limiting factor in mouse cells, since the latter have constitutively active telomerase [136]. Human, SV40-

transformed cells will undergo crisis; only sporadic cells which accumulated random mutations causing telomerase re-expression (or alternative methods to elongate chromosomal telomeres, such as ALT) will escape crisis to form an immortal cell line.

However, the host's rate of viral replication is not the sole determinant of post-infection cell susceptibility to SV40 transformation; the cellular environment appears to play an additional role. As an example, HM transfected with replication-defective SV40 genomes are still more prone to focus formation than fibroblasts (about 100 fold; [27]). Intriguingly, transfection of the aforementioned SV40 genomes synergize with crocidolite asbestos fibers (which are linked to malignant mesothelioma pathogenesis) in promoting focus formation [27]. Therefore, the SV40 transformation/immortalization potential appears to be cell type-specific, while also dependent on additional synergistic factors (e.g., crocidolite asbestos exposure). The minimal genetic elements which transform primary fibroblasts into transplantable tumors are the SV40 Tag and tag, an oncogenic ras allele, and active telomerase [137]. It is possible HM may require less genetic manipulation to become fully oncogenic.

Animal Models of SV40-Mediated Oncogenicity

SV40 is highly oncogenic in hamsters [138]. The type of tumors induced by SV40 depends on the route of viral administration. SV40 induces sarcomas if injected subcutaneously, while injection of SV40 in the brain causes ependymomas [139]. SV40 injected intravenously causes lymphomas, bone tumors, sarcomas, and leukemias [140]. The SV40 classification as polyomavirus lies in its ability to induce tumors of different histological origins. In hamsters, SV40 induces mesothelioma quite efficiently. About 60% of hamsters injected intracardially (exposing all organs to SV40) and 100% of hamsters injected intrapleurally develop mesothelioma with a short latency after viral inoculation [141]. This unusual capability of SV40 to induce mesothelioma in hamsters mirrors the high susceptibility of human and hamster mesothelial cells to SV40-mediated cell transformation in vitro.

In the hamster, SV40 is a co-carcinogen with oncogenic environmental fibers (asbestos). SV40 mutants not expressing tag fail to cause mesotheliomas in hamsters [141]. Crocidolite asbestos (the best-known pathogenic factor for mesothelioma) injected in the pleura and in the peritoneum caused mesotheliomas in about 20% of animals with a latency of 40-45 weeks. Injection of both asbestos and an SV40 strain that does not express tag caused mesothelioma in about 90% of hamsters, and these tumors arose with a much shorter latency (30-35 weeks) than those developed in hamsters injected with asbestos only [142]. SV40 and asbestos seem to synergize through the ERK/AP-1/metalloproteases axis to promote cell growth and mobility [142]. Overall, the data obtained in hamsters mirror the in vitro HM data exposed to asbestos and tag deletion mutants of SV40. Therefore, there is both in vitro and in vivo evidence SV40 can be a co-carcinogen with asbestos. A co-carcinogenic effect between a DNA oncogenic virus and an environmental agent is not unprecedented, since there is well-documented cooperation in hepatocellular carcinoma formation between hepatitis B virus and dietary aflatoxin in South-Eastern Asia [143].

Transgenic mice also underscore the high oncogenic potential of the SV40 oncogenes. Again, the type of tumor(s) mice will develop depends on the promoter regulating expression of Tag and tag (although tumor-developing transgenic mice expressing Tag alone have also been described). Mice harboring the SV40 oncogenes under the control of the SV40 regulatory region will develop choroids plexus tumors 3-5 months after birth [144]. Deletions of the so-called 72 base pair repeat in the SV40 enhancer, causing peripheral neuropathy, liver, and pancreatic tumors [145]. This different tissue tropism of SV40 oncogenes under the control of different SV40 regulatory regions may also be important in human cancer. In fact, SV40 genomes recovered from brain tumors have preferentially an "archetypal" organization of the regulatory regions, while SV40 in mesotheliomas has mostly a "wild type" regulatory region.

It should be underscored that Tag expressed under the control of tissue specific-promoters has been used to generate metastatic cancer mouse models of virtually all organs. A detailed list of all transgenic mouse models is beyond the scope of this review. Although Tag (alone or in combination with tag) has been the most effective oncogene to drive tumorigenesis in transgenic mice, there are episodic reports indicating that tag alone may be sufficient to induce neoplasia or metaplasia/dysplasia in transgenic mice. For example, pregnant mice expressing tag under the whey acidic protein promoter showed inhibition of mammary gland differentiation. About 10% of these mice developed breast tumors (although with a long latency of 10 to 17 months; [146].

The oncogenicity of SV40-transformed human cells in immunocompromized mice deserves some discussion. Generally, SV40-transformed human cells have repeatedly failed to induce tumors upon injection into immune-deficient mice. Some reports described the formation of tumor nodules at the site of injection in mice [147] and in human volunteers [148]. Nevertheless, these tumor nodules regressed with time. Different hypotheses have been proposed to account for tumor regression. Some suggested SV40-transformed human cells express viral antigens that, in nude mice, may be targeted by NK cells and macrophages [149]. Several studies, however, suggested the immune system did not have a significant role in preventing SV40-transformed human cells from forming tumors in mice [150-152]. This latter interpretation was supported by the observation that SV40-transformed human fibroblasts can become oncogenic upon transfection with oncogenes [153] or after exposure to chemical carcinogens [154]. Moreover, SV40-transformed human fibroblasts accumulate mutations during passage in tissue culture, which have occasionally led to the emergence of sub-clones which are oncogenic in mice [155,156]. These data suggest the role of the host's natural immune response (at least the innate immune system) is less important than the true oncogenic phenotype of SV40-transformed cells. Some have suggested that senescence, rather than immune rejection, is responsible for the "spontaneous" regression of tumor nodules arising following the injection of SV40-transformed human fibroblasts [157]. This hypothesis appears more likely from the results obtained with cotransfection of SV40 oncogenes, ras, and telomerase [137]. It is possible SV40-transformed human cells cannot overcome senescence in vivo unless additional factors play a role. In our hands, oncogenicity of SV40-transformed human cells is highly cell type-dependent, and the cell injection route in immunocompromized mice critically affects it.

Animal models of SV40 oncogenicity (alongside in vitro studies) have introduced a stimulating concept in SV40-mediated malignant transformation: the so-called "hit and run" mechanism [158]. According to this model, SV40 may be necessary in promoting tumor formation. During tumor development, a fraction of the cell population may acquire additional mutations that ultimately, may render the functions of the SV40 oncoproteins disposable. In such a situation, it is possible the SV40-containing cancer cells may undergo a negative selection. This is plausible since cells expressing the immunogenic SV40 oncoproteins, may be more easily targeted by the immune system, as compared to other cells in the tumor population which do not express these proteins. Also, Tag and tag would essentially become a metabolic burden for cells no longer requiring them for malignant phenotype maintenance. This model could explain why SV40 containing mesotheliomas express very low amounts of Tag [159]. Some in vivo SV40-driven tumor models support the "hit and run" mechanism. Transgenic mice expressing Tag under the control of inducible promoters show hyperplasia of the salivary glands [160], or liver, stomach, pancreas, and kidney malignancies [161]. These proliferative defects are dependent on Tag expression, since suppression of this oncogene's expression causes phenotypical reversion. But this Tag-dependency is temporarily limited, since delayed suppression of Tag expression leads to the development of malignancies no longer Tag-driven, becoming malignancies in which tag expression is disposable. Although these observations are very intriguing, there is no clear indication that the "hit and run" SV40-mediated model of viral oncogenesis occurs in humans under natural conditions.

SV40 and Humans Cancers

The prevalence of SV40 in humans is not known [162]. SV40 infection has been detected in both individuals who received SV40-contaminated vaccines, as well as those born after 1963 who did not receive the vaccine (reviewed in [9]). Because SV40's mode of transmission is uncertain, it is difficult to determine whether person-to-person transmission has occurred within the population, or if other sources of SV40 exposure exist, independent of the poliovirus vaccine [163,164]. Whatever the exposure's origin, some reports, including those showing SV40-reactive antibodies in serum of people unexposed to contaminated vaccine [3,165,166], suggest this transmission is currently occurring in human populations. Again, some other studies report a low reactivity of human sera to SV40 [167,168] or lack of evidence for prevalence of SV40 infection in humans [169].

One of the major concerns of the large-scale SV40 exposure in humans is its potential association with cancer development. The lack of epidemiological clarity, which could better define exposed and control individuals, has made it difficult to assess the cancer risk of SV40 exposed individuals. Based on the ill-defined SV40 exposure data in the human population, the Institute of Medicine Immunization and Safety Review Committee recently found epidemiological studies of cancer rates in people exposed to SV40-contaminated vaccines are inadequate to evaluate a causal relationship [170]. The Committee recommended the development and use of more specific and sensitive serological tests, standardized techniques for SV40 detection, and the study of SV40 transmissibility in humans. Even though epidemiological data is inconclusive, evidence still demonstrates SV40's capability to transform rodent and human cells in culture, and to behave oncogenically when administered in high doses to rodents, making the SV40/cancer association very controversial.

Many epidemiological studies published to date have reported the presence of SV40 DNA sequences or SV40 T antigen in several tumor types, including: brain tumors, osteosarcoma, non-Hodgkin's lymphoma, and mesothelioma [171-185], leading some investigators to propose SV40 infection may have a role in the development of these cancers. Of notice is that the four tumor types in which SV40 DNA is most frequently found in humans are also the most common tumors arising in SV40-exposed hamsters [186]. Other investigators, on the other hand, have not been able to confirm the presence of SV40 sequences, or have rarely detected it in these cancers [187](reviewed in [8]). More recently, reports of evidence for SV40's role in other human tumors have also been published. In a report from Hachana et al, 2008 concerning 109 invasive breast ductal carcinomas, the authors demonstrated the presence of SV40 in these tumors, and provided data supporting a role for this virus in their pathogenesis [188]. A second report also detected SV40 sequences (14.1%) in 78 fresh lung carcinomas tissues [189] adding to the number of tumors in which SV40 may play a role in carcinogenesis.

Limitations in the Field

Epidemiological Limitations

To date, some problems exist in the accumulated epidemiological data on the association of SV40 exposure and cancer risk, making it difficult to extract any valid conclusion from the existing studies.

First, most of the studies feature a small number of cases, or lack well-defined controls. Even though the definition of control individuals in epidemiological studies is a major problem in the association of SV40 exposure with cancer risk, the problem of small number of cases per study was recently addressed in a meta-analysis by Vilchez and collegues [190]. The meta-analysis included reports from 1975 to 2002 on only those with original studies of patients with primary brain tumor, bone cancers, malignant mesothelioma or non-Hodgkin's lymphoma. The meta-analysis also included those reports where: 1) the investigation of SV40 was performed on primary cancer specimens, 2) the study included a control group, and 3) the same techniques of detection were used in both controls and cases. The meta-analysis showed that in 13 studies, specimens from patients with brain tumors (661 tumors and 482 control samples) were almost four times more likely to show evidence of SV40 infection than those from controls (Odd Ratio= 3.9; 95%CI: 2.6-5.8). The association was seventeen times stronger for 15 mesothelioma papers (528 tumors and 468 control samples) (Odd Ratio=17, 95%CI: 10-18) and twenty five times more likely on bone cancer were data was derived from four studies (303 tumors and 121 control samples) (OR=25; 95% CI: 6.8-88). SV40 DNA was also more frequent in three studies (301 tumors and 578 controls samples) of samples from patients with non-Hodgkin's lymphoma (OR=5.4; 95%CI: 3.1). The general conclusion of this meta-analysis establishes that SV40 is significantly associated with brain tumors, bone cancers, malignant mesothelioma, and non-Hodgkin's lymphoma [190].

Another Medline search-based ecological analysis was conducted by Leithner et al, 2006 [191] covering the period from January 1969 to August 2005 for reports on the detection of SV40 DNA in human tissue samples. Their analysis showed that pleural cancer mortality in males, but not females, correlated with the extent of asbestos exposure 25-30 years earlier. In contrast, neither the presence of SV40 DNA in tumor samples, nor status of vaccination exposure had any detectable influence on cancer mortality rates in males or females. Multiple new studies have been published, some showing no association between SV40 prevalence and cancer [191-193] and some others showing a significant association with hematological malignancies [194,195] and mesothelioma [196,197].

Technical Limitation

The DNA sequence of SV40 is nearly 70% identical to other polyomaviruses found in humans (JC virus and BK virus). One difficulty that has arisen, particularly in interpreting serological data, is the potential for cross-reactivity between antibodies to SV40 and antibodies to JCV and BKV. The major capside proteins of SV40, JCV, and BKV have been shown by western blot to exhibit a cross-reacting epitope [198,199]. Efforts to discriminate among all three viruses are presently underway using ELISA methods [200,201]. One of these methods, using baculoviral-expressed virus-like proteins (VLPs), has helped distinguish between antigens of these three viruses. Recent serological studies, using VLP-based technology, have been unable to demonstrate an association between SV40 seroprevalence and poliovirus vaccination or cancer incidence [169,202-208].

Recently, another major difficulty arose from the PCR-based evidence of infection by SV40. After the recommendations of an international meeting organized in 1997 by the US National Institutes of Health and the Food and Drug Administration, two multi-center studies of SV40 detection in mesotheliomas were set up to definitively assess the presence of SV40 in mesothelioms by DNA PCR [209,210]. Even though the results of these studies were positive for the presence of SV40, inconsistencies between laboratories were detected. A report by Lopez-Rios et al, 2004 on the relationship between SV40 and homozygous deletion of CDKN2A in mesothelioma pointed out the risk of false positives due to contamination by common laboratory plasmids containing SV40 sequences, further proposing that studies based on PCR methods require careful design to reduce this risk [211]. SV40 fragments present in many commonly used laboratory plasmids (2,533 to 2,770 and 4,100 to 4,713) and also in some of the PCR primers used in different studies, were considered high-contamination risk primer pairs. Methodological details of PCR-based analysis of human mesothelioma tissue is summarized in [212]. Many studies reporting SV40 DNA in human tumors have used high-contamination risk PCR primers (e.g. SV5/SV6 and PYVfor/PYVrev). These primers amplified a region of SV40 that could be distinguished from BK and JC viral DNA, but also included the SV40 small t antigen intron and splice sites used in laboratory plasmids [7]. Even though plasmid contamination could explain the inconsistencies found in some reports, the fact still remains of the presence of SV40 in some samples. In one of these studies using mesothelioma samples, SV40 was detected by PCR in tumor cells, but not in adjacent nonmalignant cells [213].

Recent studies suggest technical limitations may account for inconsistencies in the correlation between SV40 and human tumors. Improved ELISA methods designed to differentiate between SV40, JCV, and BKV, and upgraded PCR primers designed to reduce false-positive results attributed to laboratory contamination of SV40 DNA , have set the basis for a more standardized approach for studying the correlation between SV40 and cancer. Since SV40 can transform human cells, and because studies done today still cannot deny the presence of SV40 in some tumor tissue, we cannot still exclude the possibility of an association between SV40 and cancer.

The main conclusions that can be drawn from the evidence presented in this review are that:

1) There is little doubt about the carcinogenic potential of SV40 in human cells in vitro and in animal models, and 2) There is a strong evidence of the presence of SV40 DNA in human tissue. The inconsistencies observed between laboratories concerning the detection of SV40's presence seem mainly due to the methodological approach used, or to the geographic differences associated with the study samples. The improvement of methodologies in the field, particularly those related to techniques differentiating the three poliomaviruses found in human tissues and the refinement of PCR-based analysis of SV40 DNA sequences, could shed light on these uncertainties.

The question still remains as to whether the contaminated poliovaccine is the source of SV40 presence in human tissue, and if SV40 is associated with certain human cancers. The controversy arises from the lack of conclusive epidemiologic data. It should be underscored, however, that most of the studies conducted so far have compared groups with different temporal and ethnical distribution/composition (e.g., people born before 1955 and after 1963). Such an approach is susceptible to a plethora of confounds (including the introduction of new, possibly still unidentified carcinogens, changes in lifestyle, diet, and migration patterns) virtually impossible to take into account. We now know that contamination of poliovaccines with infectious SV40 continued well into the Seventies in the Soviet block countries [214]. This implies that until relatively recently, hundreds of millions of people have had continued exposure to SV40 through poliovaccination. The unusual prevalence of SV40 in Kazakhstan's population supports this idea [215].

Two epidemiologic studies conducted on pregnant women receiving SV40-contaminated poliovaccines as compared to women who did not had the advantage of comparing exposed versus non exposed individuals at the same time and in similar environmental and socio-economic conditions [216,217]. Strikingly, both studies showed an increased risk of pediatric cancer development in the offspring of the vaccinated mothers, as compared to those who did not receive the poliovaccine. However, these studies suffered the shortcoming of small cohorts included in the analyses.

We can safely state that conclusive epidemiologic evidence linking SV40 to human cancer pathogenesis is lacking. It is intriguing, however, that mesothelioma, a disease only anecdotically reported before the Fifties, has risen to 2,000-3,000 cases per year currently. This is undoubtedly due, at least in part, to increased exposure to carcinogenic fibers, but also coincided with the large-scale vaccination of people with SV40-contaminated poliovaccines. Since carcinogenesis is notoriously multifactorial, we cannot exclude SV40's potential contribution to the unusual rise of mesothelioma worldwide after the second half of the 20th century.

Acknowledgments

Grant NCI-K01 CA104159 (MER-N). and American Cancer Society grant RSG-05-077 (MB)

References

[1] White MK, Khalili K: Polyomaviruses and human cancer: molecular mechanisms underlying patterns of tumorigenesis. *Virology* 2004, 324:1-16.

[2] Soriano F, Shelburne CE, Gokcen M: Simian virus 40 in a human cancer. *Nature* 1974, 249:421-424.

[3] Shah KV, Palma LD, Murphy GP: The occurrence of SV40-Neutralizing antibodies in sera of patients with genitourinary carcinoma. *J Surg Oncol* 1971, 3:443-450.

[4] Shah KV, Weismann M, Murphy GP: Occurrence of simian virus 40 (SV40)-reacting antibodies in sera of some patients with prostatic cancer. *J Surg Oncol* 1972, 4:89-93.

[5] Gazdar A, Butel J, Carbone M: SV40 and human tumours: myth, association or causality? *Nat Rev Cancer* 2002, 2:957-964.

[6] Klein G, Powers A, Croce C: Association of SV40 with human tumors. *Oncogene* 2002, 21:1141-1149.

[7] Poulin DL, DeCaprio JA: Is there a role for SV40 in human cancer? *J Clin Oncol* 2006, 24:4356-4365.

[8] Shah KV: SV40 and human cancer: a review of recent data. *Int J Cancer* 2007, 120:215-223.

[9] Vilchez RA, Butel JS: Emergent human pathogen simian virus 40 and its role in cancer. *Clin Microbiol Rev* 2004, 17:495-508, table of contents.

[10] Butel JS, Lednicky JA: Cell and molecular biology of simian virus 40: implications for human infections and disease. *J Natl Cancer Inst* 1999, 91:119-134.

[11] Sweet BH, Hilleman MR: The vacuolating virus, S.V. 40. *Proc Soc Exp Biol Med* 1960, 105:420-427.

[12] Kosukegawa A, Arisaka F, Takayama M, Yajima H, Kaidow A, Handa H: Purification and characterization of virus-like particles and pentamers produced by the expression of SV40 capsid proteins in insect cells. *Biochim Biophys Acta* 1996, 1290:37-45.

[13] Sandalon Z, Oppenheim A: Self-assembly and protein-protein interactions between the SV40 capsid proteins produced in insect cells. *Virology* 1997, 237:414-421.

[14] Kawano MA, Inoue T, Tsukamoto H, Takaya T, Enomoto T, Takahashi RU, Yokoyama N, Yamamoto N, Nakanishi A, Imai T, et al: The VP2/VP3 minor capsid protein of simian virus 40 promotes the in vitro assembly of the major capsid protein VP1 into particles. *J Biol Chem* 2006, 281:10164-10173.

[15] Salunke DM, Caspar DL, Garcea RL: Self-assembly of purified polyomavirus capsid protein VP1. *Cell* 1986, 46:895-904.

[16] Cordelier P, Bienvenu C, Lulka H, Marrache F, Bouisson M, Openheim A, Strayer DS, Vaysse N, Pradayrol L, Buscail L: Replication-deficient rSV40 mediate pancreatic gene transfer and long-term inhibition of tumor growth. *Cancer Gene Ther* 2007, 14:19-29.

[17] Eid L, Bromberg Z, El-Latif MA, Zeira E, Oppenheim A, Weiss YG: Simian virus 40 vectors for pulmonary gene therapy. *Respir Res* 2007, 8:74.

[18] Oppenheim A, Peleg A, Fibach E, Rachmilewitz EA: Efficient introduction of plasmid DNA into human hemopoietic cells by encapsidation in simian virus 40 pseudovirions. *Proc Natl Acad Sci U S A* 1986, 83:6925-6929.

[19] Fiers W, Contreras R, Haegemann G, Rogiers R, Van de Voorde A, Van Heuverswyn H, Van Herreweghe J, Volckaert G, Ysebaert M: Complete nucleotide sequence of SV40 DNA. *Nature* 1978, 273:113-120.

[20] Muller U, Zentgraf H, Eicken I, Keller W: Higher order structure of simian virus 40 chromatin. *Science* 1978, 201:406-415.

[21] Nedospasov SA, Bakayev VV, Georgiev GP: Chromosome of the mature virion of simian virus 40 contains H1 histone. *Nucleic Acids Res* 1978, 5:2847-2860.

[22] Murakami Y, Wobbe CR, Weissbach L, Dean FB, Hurwitz J: Role of DNA polymerase alpha and DNA primase in simian virus 40 DNA replication in vitro. *Proc Natl Acad Sci U S A* 1986, 83:2869-2873.

[23] Eki T, Enomoto T, Masutani C, Miyajima A, Takada R, Murakami Y, Ohno T, Hanaoka F, Ui M: Mouse DNA primase plays the principal role in determination of permissiveness for polyomavirus DNA replication. *J Virol* 1991, 65:4874-4881.

[24] Schneider C, Weisshart K, Guarino LA, Dornreiter I, Fanning E: Species-specific functional interactions of DNA polymerase alpha-primase with simian virus 40 (SV40) T antigen require SV40 origin DNA. *Mol Cell Biol* 1994, 14:3176-3185.

[25] Clayson ET, Brando LV, Compans RW: Release of simian virus 40 virions from epithelial cells is polarized and occurs without cell lysis. *J Virol* 1989, 63:2278-2288.

[26] Carp RI, Gilden RV: A comparison of the replication cycles of simian virus 40 in human diploid and African green monkey kidney cells. *Virology* 1966, 28:150-162.

[27] Bocchetta M, Di Resta I, Powers A, Fresco R, Tosolini A, Testa J, Pass H, Rizzo P, Carbone M: Human mesothelial cells are unusually susceptible to simian virus 40-mediated transformation and asbestos cocarcinogenicity. *Proc Natl Acad Sci USA* 2000, 97:10214-10219.

[28] Atwood WJ, Norkin LC: Class I major histocompatibility proteins as cell surface receptors for simian virus 40. *J Virol* 1989, 63:4474-4477.

[29] Breau WC, Atwood WJ, Norkin LC: Class I major histocompatibility proteins are an essential component of the simian virus 40 receptor. *J Virol* 1992, 66:2037-2045.

[30] Basak S, Turner H, Compans RW: Expression of SV40 receptors on apical surfaces of polarized epithelial cells. *Virology* 1992, 190:393-402.

[31] Kartenbeck J, Stukenbrok H, Helenius A: Endocytosis of simian virus 40 into the endoplasmic reticulum. *J Cell Biol* 1989, 109:2721-2729.

[32] Anderson HA, Chen Y, Norkin LC: Bound simian virus 40 translocates to caveolin-enriched membrane domains, and its entry is inhibited by drugs that selectively disrupt caveolae. *Mol Biol Cell* 1996, 7:1825-1834.

[33] Anderson HA, Chen Y, Norkin LC: MHC class I molecules are enriched in caveolae but do not enter with simian virus 40. *J Gen Virol* 1998, 79 (Pt 6):1469-1477.

[34] Norkin LC, Anderson HA, Wolfrom SA, Oppenheim A: Caveolar endocytosis of simian virus 40 is followed by brefeldin A-sensitive transport to the endoplasmic reticulum, where the virus disassembles. *J Virol* 2002, 76:5156-5166.

[35] Schelhaas M, Malmstrom J, Pelkmans L, Haugstetter J, Ellgaard L, Grunewald K, Helenius A: Simian Virus 40 depends on ER protein folding and quality control factors for entry into host cells. *Cell* 2007, 131:516-529.

[36] Damm EM, Pelkmans L, Kartenbeck J, Mezzacasa A, Kurzchalia T, Helenius A: Clathrin- and caveolin-1-independent endocytosis: entry of simian virus 40 into cells devoid of caveolae. *J Cell Biol* 2005, 168:477-488.

[37] Yamada M, Kasamatsu H: Role of nuclear pore complex in simian virus 40 nuclear targeting. *J Virol* 1993, 67:119-130.

[38] Nakanishi A, Clever J, Yamada M, Li PP, Kasamatsu H: Association with capsid proteins promotes nuclear targeting of simian virus 40 DNA. *Proc Natl Acad Sci U S A* 1996, 93:96-100.

[39] Nakanishi A, Itoh N, Li PP, Handa H, Liddington RC, Kasamatsu H: Minor capsid proteins of simian virus 40 are dispensable for nucleocapsid assembly and cell entry but are required for nuclear entry of the viral genome. *J Virol* 2007, 81:3778-3785.

[40] Dean DA, Kasamatsu H: Signal- and energy-dependent nuclear transport of SV40 Vp3 by isolated nuclei. Establishment of a filtration assay for nuclear protein import. *J Biol Chem* 1994, 269:4910-4916.

[41] Gordon-Shaag A, Ben-Nun-Shaul O, Roitman V, Yosef Y, Oppenheim A: Cellular transcription factor Sp1 recruits simian virus 40 capsid proteins to the viral packaging signal, ses. *J Virol* 2002, 76:5915-5924.

[42] Gordon-Shaag A, Ben-Nun-Shaul O, Kasamatsu H, Oppenheim AB, Oppenheim A: The SV40 capsid protein VP3 cooperates with the cellular transcription factor Sp1 in DNA-binding and in regulating viral promoter activity. *J Mol Biol* 1998, 275:187-195.

[43] Wiley SR, Kraus RJ, Zuo F, Murray EE, Loritz K, Mertz JE: SV40 early-to-late switch involves titration of cellular transcriptional repressors. *Genes Dev* 1993, 7:2206-2219.

[44] Clemens MJ: PKR--a protein kinase regulated by double-stranded RNA. *Int J Biochem Cell Biol* 1997, 29:945-949.

[45] Cole C: *Fields Virology.* Philadelphia: Lippincott-Raven; 1996.

[46] Berk AJ, Sharp PA: Spliced early mRNAs of simian virus 40. *Proc Natl Acad Sci U S A* 1978, 75:1274-1278.

[47] Paucha E, Smith AE: The sequences between 0.59 and 0.54 map units on SV40 DNA code for the unique region of small t antigen. *Cell* 1978, 15:1011-1020.

[48] May E, Lasne C, Prives C, Borde J, May P: Study of the functional activities concomitantly retained by the 115,000 Mr super T antigen, an evolutionary variant of simian virus 40 large T antigen expressed in transformed rat cells. *J Virol* 1983, 45:901-913.

[49] Boyapati A, Wilson M, Yu J, Rundell K: SV40 17KT antigen complements dnaj mutations in large T antigen to restore transformation of primary human fibroblasts. *Virology* 2003, 315:148-158.

[50] Sedman SA, Good PJ, Mertz JE: Leader-encoded open reading frames modulate both the absolute and relative rates of synthesis of the virion proteins of simian virus 40. *J Virol* 1989, 63:3884-3893.
[51] Tjian R: Regulation of viral transcription and DNA replication by the SV40 large T antigen. *Curr Top Microbiol Immunol* 1981, 93:5-24.
[52] Wobbe CR, Dean F, Weissbach L, Hurwitz J: In vitro replication of duplex circular DNA containing the simian virus 40 DNA origin site. *Proc Natl Acad Sci U S A* 1985, 82:5710-5714.
[53] Stahl H, Droge P, Knippers R: DNA helicase activity of SV40 large tumor antigen. *Embo J* 1986, 5:1939-1944.
[54] Gannon JV, Lane DP: p53 and DNA polymerase alpha compete for binding to SV40 T antigen. *Nature* 1987, 329:456-458.
[55] Damania B, Alwine JC: TAF-like function of SV40 large T antigen. *Genes Dev* 1996, 10:1369-1381.
[56] Avantaggiati ML, Carbone M, Graessmann A, Nakatani Y, Howard B, Levine AS: The SV40 large T antigen and adenovirus E1a oncoproteins interact with distinct isoforms of the transcriptional co-activator, p300. *Embo J* 1996, 15:2236-2248.
[57] DeCaprio JA, Ludlow JW, Figge J, Shew JY, Huang CM, Lee WH, Marsilio E, Paucha E, Livingston DM: SV40 large tumor antigen forms a specific complex with the product of the retinoblastoma susceptibility gene. *Cell* 1988, 54:275-283.
[58] Sawai ET, Butel JS: Association of a cellular heat shock protein with simian virus 40 large T antigen in transformed cells. *J Virol* 1989, 63:3961-3973.
[59] Miyata Y, Yahara I: p53-independent association between SV40 large T antigen and the major cytosolic heat shock protein, HSP90. *Oncogene* 2000, 19:1477-1484.
[60] Kasper JS, Kuwabara H, Arai T, Ali SH, DeCaprio JA: Simian virus 40 large T antigen's association with the CUL7 SCF complex contributes to cellular transformation. *J Virol* 2005, 79:11685-11692.
[61] Welcker M, Clurman BE: The SV40 large T antigen contains a decoy phosphodegron that mediates its interactions with Fbw7/hCdc4. *J Biol Chem* 2005, 280:7654-7658.
[62] Wu X, Avni D, Chiba T, Yan F, Zhao Q, Lin Y, Heng H, Livingston D: SV40 T antigen interacts with Nbs1 to disrupt DNA replication control. *Genes Dev* 2004, 18:1305-1316.
[63] Pinhasi-Kimhi O, Michalovitz D, Ben-Zeev A, Oren M: Specific interaction between the p53 cellular tumour antigen and major heat shock proteins. *Nature* 1986, 320:182-184.
[64] Kojima T, Ikawa Y, Katoh I: Analysis of molecular interactions of the p53-family p51(p63) gene products in a yeast two-hybrid system: homotypic and heterotypic interactions and association with p53-regulatory factors. *Biochem Biophys Res Commun* 2001, 281:1170-1175.
[65] Bocchetta M, Miele L, Pass HI, Carbone M: Notch-1 induction, a novel activity of SV40 required for growth of SV40-transformed human mesothelial cells. *Oncogene* 2003, 22:81-89.

[66] Bocchetta M, Eliasz S, De Marco MA, Rudzinski J, Zhang L, Carbone M: The SV40 large T antigen-p53 complexes bind and activate the insulin-like growth factor-I promoter stimulating cell growth. *Cancer Res* 2008, 68:1022-1029.

[67] Porcu P, Grana X, Li S, Swantek J, De Luca A, Giordano A, Baserga R: An E2F binding sequence negatively regulates the response of the insulin-like growth factor 1 (IGF-I) promoter to simian virus 40T antigen and to serum. *Oncogene* 1994, 9:2125-2134.

[68] Chen H, Campisi J, Padmanabhan R: SV40 large T antigen transactivates the human cdc2 promoter by inducing a CCAAT box binding factor. *J Biol Chem* 1996, 271:13959-13967.

[69] Cacciotti P, Libener R, Betta P, Martini F, Porta C, Procopio A, Strizzi L, Penengo L, Tognon M, Mutti L, Gaudino G: SV40 replication in human mesothelial cells induces HGF/Met receptor activation: a model for viral-related carcinogenesis of human malignant mesothelioma. *Proc Natl Acad Sci USA* 2001, 98:12032-12037.

[70] Pallas DC, Shahrik LK, Martin BL, Jaspers S, Miller TB, Brautigan DL, Roberts TM: Polyoma small and middle T antigens and SV40 small t antigen form stable complexes with protein phosphatase 2A. *Cell* 1990, 60:167-176.

[71] Wheat WH, Roesler WJ, Klemm DJ: Simian virus 40 small tumor antigen inhibits dephosphorylation of protein kinase A-phosphorylated CREB and regulates CREB transcriptional stimulation. *Mol Cell Biol* 1994, 14:5881-5890.

[72] Watanabe G, Howe A, Lee RJ, Albanese C, Shu IW, Karnezis AN, Zon L, Kyriakis J, Rundell K, Pestell RG: Induction of cyclin D1 by simian virus 40 small tumor antigen. *Proc Natl Acad Sci U S A* 1996, 93:12861-12866.

[73] Skoczylas C, Henglein B, Rundell K: PP2A-dependent transactivation of the cyclin A promoter by SV40 ST is mediated by a cell cycle-regulated E2F site. *Virology* 2005, 332:596-601.

[74] Sontag E, Sontag JM, Garcia A: Protein phosphatase 2A is a critical regulator of protein kinase C zeta signaling targeted by SV40 small t to promote cell growth and NF-kappaB activation. *Embo J* 1997, 16:5662-5671.

[75] Cicchillitti L, Fasanaro P, Biglioli P, Capogrossi MC, Martelli F: Oxidative stress induces protein phosphatase 2A-dependent dephosphorylation of the pocket proteins pRb, p107, and p130. *J Biol Chem* 2003, 278:19509-19517.

[76] Ma J, Arnold HK, Lilly MB, Sears RC, Kraft AS: Negative regulation of Pim-1 protein kinase levels by the B56beta subunit of PP2A. *Oncogene* 2007, 26:5145-5153.

[77] Garriga J, Jayaraman AL, Limon A, Jayadeva G, Sotillo E, Truongcao M, Patsialou A, Wadzinski BE, Grana X: A dynamic equilibrium between CDKs and PP2A modulates phosphorylation of pRB, p107 and p130. *Cell Cycle* 2004, 3:1320-1330.

[78] Yuan H, Veldman T, Rundell K, Schlegel R: Simian virus 40 small tumor antigen activates AKT and telomerase and induces anchorage-independent growth of human epithelial cells. *J Virol* 2002, 76:10685-10691.

[79] Hui L, Rodrik V, Pielak RM, Knirr S, Zheng Y, Foster DA: mTOR-dependent suppression of protein phosphatase 2A is critical for phospholipase D survival signals in human breast cancer cells. *J Biol Chem* 2005, 280:35829-35835.

[80] Xin M, Deng X: Protein phosphatase 2A enhances the proapoptotic function of Bax through dephosphorylation. *J Biol Chem* 2006, 281:18859-18867.

[81] Keller JM, Alwine JC: Activation of the SV40 late promoter: direct effects of T antigen in the absence of viral DNA replication. *Cell* 1984, 36:381-389.

[82] Zuo F, Mertz JE: Simian virus 40 late gene expression is regulated by members of the steroid/thyroid hormone receptor superfamily. *Proc Natl Acad Sci U S A* 1995, 92:8586-8590.

[83] Desvergne B, Favez T: The major transcription initiation site of the SV40 late promoter is a potent thyroid hormone response element. *Nucleic Acids Res* 1997, 25:1774-1781.

[84] Lee HJ, Chang C: Identification of human TR2 orphan receptor response element in the transcriptional initiation site of the simian virus 40 major late promoter. *J Biol Chem* 1995, 270:5434-5440.

[85] Farrell ML, Mertz JE: Cell type-specific replication of simian virus 40 conferred by hormone response elements in the late promoter. *J Virol* 2002, 76:6762-6770.

[86] Laffin J, Fogleman D, Lehman JM: Correlation of DNA content, p53, T antigen, and V antigen in simian virus 40-infected human diploid cells. *Cytometry* 1989, 10:205-213.

[87] Okubo E, Lehman JM, Friedrich TD: Negative regulation of mitotic promoting factor by the checkpoint kinase chk1 in simian virus 40 lytic infection. *J Virol* 2003, 77:1257-1267.

[88] Sullivan CS, Grundhoff AT, Tevethia S, Pipas JM, Ganem D: SV40-encoded microRNAs regulate viral gene expression and reduce susceptibility to cytotoxic T cells. *Nature* 2005, 435:682-686.

[89] Gordon-Shaag A, Yosef Y, Abd El-Latif M, Oppenheim A: The abundant nuclear enzyme PARP participates in the life cycle of simian virus 40 and is stimulated by minor capsid protein VP3. *J Virol* 2003, 77:4273-4282.

[90] Daniels R, Rusan NM, Wilbuer AK, Norkin LC, Wadsworth P, Hebert DN: Simian virus 40 late proteins possess lytic properties that render them capable of permeabilizing cellular membranes. *J Virol* 2006, 80:6575-6587.

[91] Resnick J, Shenk T: Simian virus 40 agnoprotein facilitates normal nuclear location of the major capsid polypeptide and cell-to-cell spread of virus. *J Virol* 1986, 60:1098-1106.

[92] Hou-Jong MH, Larsen SH, Roman A: Role of the agnoprotein in regulation of simian virus 40 replication and maturation pathways. *J Virol* 1987, 61:937-939.

[93] Khalili K, White MK, Sawa H, Nagashima K, Safak M: The agnoprotein of polyomaviruses: a multifunctional auxiliary protein. *J Cell Physiol* 2005, 204:1-7.

[94] Del Valle L, Gordon J, Enam S, Delbue S, Croul S, Abraham S, Radhakrishnan S, Assimakopoulou M, Katsetos CD, Khalili K: Expression of human neurotropic polyomavirus JCV late gene product agnoprotein in human medulloblastoma. *J Natl Cancer Inst* 2002, 94:267-273.

[95] Ludlow JW, Shon J, Pipas JM, Livingston DM, DeCaprio JA: The retinoblastoma susceptibility gene product undergoes cell cycle-dependent dephosphorylation and binding to and release from SV40 large T. *Cell* 1990, 60:387-396.

[96] Royzman I, Whittaker AJ, Orr-Weaver TL: Mutations in Drosophila DP and E2F distinguish G1-S progression from an associated transcriptional program. *Genes Dev* 1997, 11:1999-2011.

[97] Chang YC, Illenye S, Heintz NH: Cooperation of E2F-p130 and Sp1-pRb complexes in repression of the Chinese hamster dhfr gene. *Mol Cell Biol* 2001, 21:1121-1131.

[98] Chan HM, Smith L, La Thangue NB: Role of LXCXE motif-dependent interactions in the activity of the retinoblastoma protein. *Oncogene* 2001, 20:6152-6163.

[99] Gray SG, Iglesias AH, Lizcano F, Villanueva R, Camelo S, Jingu H, Teh BT, Koibuchi N, Chin WW, Kokkotou E, Dangond F: Functional characterization of JMJD2A, a histone deacetylase- and retinoblastoma-binding protein. *J Biol Chem* 2005, 280:28507-28518.

[100] Chan HM, Krstic-Demonacos M, Smith L, Demonacos C, La Thangue NB: Acetylation control of the retinoblastoma tumour-suppressor protein. *Nat Cell Biol* 2001, 3:667-674.

[101] Milton AH, Khaire N, Ingram L, O'Donnell AJ, La Thangue NB: 14-3-3 proteins integrate E2F activity with the DNA damage response. *Embo J* 2006, 25:1046-1057.

[102] Bargonetti J, Reynisdottir I, Friedman PN, Prives C: Site-specific binding of wild-type p53 to cellular DNA is inhibited by SV40 T antigen and mutant p53. *Genes Dev* 1992, 6:1886-1898.

[103] Jiang D, Srinivasan A, Lozano G, Robbins PD: SV40 T antigen abrogates p53-mediated transcriptional activity. *Oncogene* 1993, 8:2805-2812.

[104] Segawa K, Minowa A, Sugasawa K, Takano T, Hanaoka F: Abrogation of p53-mediated transactivation by SV40 large T antigen. *Oncogene* 1993, 8:543-548.

[105] Deppert W, Haug M: Evidence for free and metabolically stable p53 protein in nuclear subfractions of simian virus 40-transformed cells. *Mol Cell Biol* 1986, 6:2233-2240.

[106] O'Neill FJ, Hu Y, Chen T, Carney H: Identification of p53 unbound to T-antigen in human cells transformed by simian virus 40 T-antigen. *Oncogene* 1997, 14:955-965.

[107] Long SB, Ho HY, Chen CL, Lai MD: Complex of simian virus large T antigen and p53 can bind DNA specifically. *Anticancer Res* 1995, 15:1375-1380.

[108] Sheppard HM, Corneillie SI, Espiritu C, Gatti A, Liu X: New insights into the mechanism of inhibition of p53 by simian virus 40 large T antigen. *Mol Cell Biol* 1999, 19:2746-2753.

[109] Technau A, Wolff A, Sauder C, Birkner N, Brandner G: p53 in SV40-transformed DNA-damaged human cells binds to its cognate sequence but fails to transactivate target genes. *Int J Oncol* 2001, 18:281-286.

[110] Michalovitz D, Eliyahu D, Oren M: Overproduction of protein p53 contributes to simian virus 40-mediated transformation. *Mol Cell Biol* 1986, 6:3531-3536.

[111] Deppert W, Steinmayer T, Richter W: Cooperation of SV40 large T antigen and the cellular protein p53 in maintenance of cell transformation. *Oncogene* 1989, 4:1103-1110.

[112] Herzig M, Novatchkova M, Christofori G: An unexpected role for p53 in augmenting SV40 large T antigen-mediated tumorigenesis. *Biol Chem* 1999, 380:203-211.

[113] Carbone M, Burck C, Rdzanek M, Rudzinski J, Cutrone R, Bocchetta M: Different susceptibility of human mesothelial cells to polyomavirus infection and malignant transformation. *Cancer Res* 2003, 63:6125-6129.

[114] Porcu P, Ferber A, Pietrzkowski Z, Roberts CT, Adamo M, LeRoith D, Baserga R: The growth-stimulatory effect of simian virus 40 T antigen requires the interaction of insulinlike growth factor 1 with its receptor. *Mol Cell Biol* 1992, 12:5069-5077.

[115] Sachsenmeier KF, Pipas JM: Inhibition of Rb and p53 is insufficient for SV40 T-antigen transformation. *Virology* 2001, 283:40-48.

[116] Lednicky JA, Arrington AS, Stewart AR, Dai XM, Wong C, Jafar S, Murphey-Corb M, Butel JS: Natural isolates of simian virus 40 from immunocompromised monkeys display extensive genetic heterogeneity: new implications for polyomavirus disease. *J Virol* 1998, 72:3980-3990.

[117] Berger JR: Progressive multifocal leukoencephalopathy in acquired immunodeficiency syndrome: explaining the high incidence and disproportionate frequency of the illness relative to other immunosuppressive conditions. *J Neurovirol* 2003, 9 Suppl 1:38-41.

[118] Bryan TM, Reddel RR: SV40-induced immortalization of human cells. *Crit Rev Oncog* 1994, 5:331-357.

[119] de Ronde A, Sol CJ, van Strien A, ter Schegget J, van der Noordaa J: The SV40 small t antigen is essential for the morphological transformation of human fibroblasts. *Virology* 1989, 171:260-263.

[120] Hayflick L: The Limited in Vitro Lifetime of Human Diploid Cell Strains. *Exp Cell Res* 1965, 37:614-636.

[121] Wei W, Hemmer RM, Sedivy JM: Role of p14(ARF) in replicative and induced senescence of human fibroblasts. *Mol Cell Biol* 2001, 21:6748-6757.

[122] Bodnar AG, Ouellette M, Frolkis M, Holt SE, Chiu CP, Morin GB, Harley CB, Shay JW, Lichtsteiner S, Wright WE: Extension of life-span by introduction of telomerase into normal human cells. *Science* 1998, 279:349-352.

[123] Macera-Bloch L, Houghton J, Lenahan M, Jha KK, Ozer HL: Termination of lifespan of SV40-transformed human fibroblasts in crisis is due to apoptosis. *J Cell Physiol* 2002, 190:332-344.

[124] O'Neill FJ, Carney H, Hu Y: Host range analysis of simian virus 40, BK virus and chimaeric SV40/BKV: relative expression of large T-antigen and Vp1 in infected and transformed cells. *Dev Biol Stand* 1998, 94:191-205.

[125] Croce CM, Girardi AJ, Koprowski H: Assignment of the T-antigen gene of simian virus 40 to human chromosome C-7. *Proc Natl Acad Sci U S A* 1973, 70:3617-3620.

[126] Croce CM, Huebner K, Girardi AJ, Koprowski H: Genetics of cell transformation by simian virus 40. *Cold Spring Harb Symp Quant Biol* 1975, 39 Pt 1:335-343.

[127] Mishmar D, Rahat A, Scherer SW, Nyakatura G, Hinzmann B, Kohwi Y, Mandel-Gutfroind Y, Lee JR, Drescher B, Sas DE, et al: Molecular characterization of a common fragile site (FRA7H) on human chromosome 7 by the cloning of a simian virus 40 integration site. *Proc Natl Acad Sci U S A* 1998, 95:8141-8146.

[128] Rabin M, Uhlenbeck OC, Steffensen DM, Mangel WF: Chromosomal sites of integration of simian virus 40 DNA sequences mapped by in situ hybridization in two transformed hybrid cell lines. *J Virol* 1984, 49:445-451.

[129] Croce CM: Assignment of the integration site for simian virus 40 to chromosome 17 in GM54VA, a human cell line transformed by simian virus 40. *Proc Natl Acad Sci U S A* 1977, 74:315-318.

[130] Shera KA, Shera CA, McDougall JK: Small tumor virus genomes are integrated near nuclear matrix attachment regions in transformed cells. *J Virol* 2001, 75:12339-12346.

[131] Huang KC, Yamasaki EF, Snapka RM: Maintenance of episomal SV40 genomes in GM637 human fibroblasts. *Virology* 1999, 262:457-469.

[132] Morelli C, Barbisan F, Iaccheri L, Tognon M: Simian virus 40 persistent infection in long-term immortalized human fibroblast cell lines. *J Neurovirol* 2004, 10:250-254.

[133] Mendoza SM, Konishi T, Miller CW: Integration of SV40 in human osteosarcoma DNA. *Oncogene* 1998, 17:2457-2462.

[134] Krieg P, Amtmann E, Sauer G, Lavi S, Kleinberger T, Winocour E: The integrated SV40 genome in permissive transformed monkey cells. *Virology* 1981, 108:453-461.

[135] Foddis R, De Rienzo A, Broccoli D, Bocchetta M, Stekala E, Rizzo P, Tosolini A, Grobelny JV, Jhanwar SC, Pass HI, et al: SV40 infection induces telomerase activity in human mesothelial cells. *Oncogene* 2002, 21:1434-1442.

[136] Chadeneau C, Siegel P, Harley CB, Muller WJ, Bacchetti S: Telomerase activity in normal and malignant murine tissues. *Oncogene* 1995, 11:893-898.

[137] Hahn WC, Counter CM, Lundberg AS, Beijersbergen RL, Brooks MW, Weinberg RA: Creation of human tumour cells with defined genetic elements. *Nature* 1999, 400:464-468.

[138] Eddy BE, Borman GS, Berkeley WH, Young RD: Tumors induced in hamsters by injection of rhesus monkey kidney cell extracts. *Proc Soc Exp Biol Med* 1961, 107:191-197.

[139] Kirschstein RL, Gerber P: Ependymomas produced after intracerebral inoculation of SV40 into new-born hamsters. *Nature* 1962, 195:299-300.

[140] Diamandopoulos GT: Induction of lymphocytic leukemia, lymphosarcoma, reticulum cell sarcoma, and osteogenic sarcoma in the Syrian golden hamster by oncogenic DNA simian virus 40. *J Natl Cancer Inst* 1973, 50:1347-1365.

[141] Cicala C, Pompetti F, Carbone M: SV40 induces mesotheliomas in hamsters. *Am J Pathol* 1993, 142:1524-1533.

[142] Kroczynska B, Cutrone R, Bocchetta M, Yang H, Elmishad AG, Vacek P, Ramos-Nino M, Mossman BT, Pass HI, Carbone M: Crocidolite asbestos and SV40 are cocarcinogens in human mesothelial cells and in causing mesothelioma in hamsters. *Proc Natl Acad Sci U S A* 2006, 103:14128-14133.

[143] Ming L, Thorgeirsson SS, Gail MH, Lu P, Harris CC, Wang N, Shao Y, Wu Z, Liu G, Wang X, Sun Z: Dominant role of hepatitis B virus and cofactor role of aflatoxin in hepatocarcinogenesis in Qidong, China. *Hepatology* 2002, 36:1214-1220.

[144] Palmiter RD, Chen HY, Messing A, Brinster RL: SV40 enhancer and large-T antigen are instrumental in development of choroid plexus tumours in transgenic mice. *Nature* 1985, 316:457-460.

[145] Messing A, Chen HY, Palmiter RD, Brinster RL: Peripheral neuropathies, hepatocellular carcinomas and islet cell adenomas in transgenic mice. *Nature* 1985, 316:461-463.

[146] Goetz F, Tzeng YJ, Guhl E, Merker J, Graessmann M, Graessmann A: The SV40 small t-antigen prevents mammary gland differentiation and induces breast cancer formation in transgenic mice; truncated large T-antigen molecules harboring the intact p53 and pRb binding region do not have this effect. *Oncogene* 2001, 20:2325-2332.

[147] Croce CM, Aden D, Koprowski H: Somatic cell hybrids between mouse peritoneal macrophages and simian-virus-40-transformed human cells: II. Presence of human chromosome 7 carrying simin virus 40 genome in cells of tumors induced by hybrid cells. *Proc Natl Acad Sci U S A* 1975, 72:1397-1400.

[148] Jensen FC, Girardi AJ, Gilden RV, Koprowski H: Infection of Human and Simian Tissue Cultures with Rous Sarcoma Virus. *Proc Natl Acad Sci U S A* 1964, 52:53-59.

[149] Choi KH, Tevethia SS, Shin S: Tumor formation by SV40-transformed human cells in nude mice: the role of SV40 T antigens. *Cytogenet Cell Genet* 1983, 36:633-640.

[150] Stiles CD, Desmond W, Jr., Sato G, Saier MH, Jr.: Failure of human cells transformed by simian virus 40 to form tumors in athymic nude mice. *Proc Natl Acad Sci U S A* 1975, 72:4971-4975.

[151] Kahn P, Topp WC, Shin S: Tumorigenicity of SV40-transformed human and monkey cells in immunodeficient mice. *Virology* 1983, 126:348-360.

[152] Sager R, Tanaka K, Lau CC, Ebina Y, Anisowicz A: Resistance of human cells to tumorigenesis induced by cloned transforming genes. *Proc Natl Acad Sci U S A* 1983, 80:7601-7605.

[153] Reddel RR, De Silva R, Duncan EL, Rogan EM, Whitaker NJ, Zahra DG, Ke Y, McMenamin MG, Gerwin BI, Harris CC: SV40-induced immortalization and ras-transformation of human bronchial epithelial cells. *Int J Cancer* 1995, 61:199-205.

[154] Sasaki K, Mironov N, Yilmaz A, Lahm H, Odartchenko N, Yamasaki H: Malignant transformation of simian virus 40-immortalized human milk epithelial cells by chemical carcinogenesis accompanied by loss of heterozygosity on chromosome 1 but not microsatellite instability. *Mol Carcinog* 1998, 23:20-24.

[155] Reddel RR, Salghetti SE, Willey JC, Ohnuki Y, Ke Y, Gerwin BI, Lechner JF, Harris CC: Development of tumorigenicity in simian virus 40-immortalized human bronchial epithelial cell lines. *Cancer Res* 1993, 53:985-991.

[156] Yilmaz A, Gaide AC, Sordat B, Borbenyi Z, Lahm H, Imam A, Schreyer M, Odartchenko N: Malignant progression of SV40-immortalised human milk epithelial cells. *Br J Cancer* 1993, 68:868-873.

[157] Sager R: Senescence as a mode of tumor suppression. *Environ Health Perspect* 1991, 93:59-62.

[158] Wiman KG, Klein G: An old acquaintance resurfaces in human mesothelioma. *Nat Med* 1997, 3:839-840.

[159] Waheed I, Guo ZS, Chen GA, Weiser TS, Nguyen DM, Schrump DS: Antisense to SV40 early gene region induces growth arrest and apoptosis in T-antigen-positive human pleural mesothelioma cells. *Cancer Res* 1999, 59:6068-6073.

[160] Ewald D, Li M, Efrat S, Auer G, Wall RJ, Furth PA, Hennighausen L: Time-sensitive reversal of hyperplasia in transgenic mice expressing SV40 T antigen. *Science* 1996, 273:1384-1386.

[161] Sepulveda AR, Finegold MJ, Smith B, Slagle BL, DeMayo JL, Shen RF, Woo SL, Butel JS: Development of a transgenic mouse system for the analysis of stages in liver carcinogenesis using tissue-specific expression of SV40 large T-antigen controlled by regulatory elements of the human alpha-1-antitrypsin gene. *Cancer Res* 1989, 49:6108-6117.

[162] Vilchez RA, Kozinetz CA, Butel JS: Conventional epidemiology and the link between SV40 and human cancers. *Lancet Oncol* 2003, 4:188-191.

[163] Minor P, Pipkin PA, Cutler K, Dunn G: Natural infection and transmission of SV40. *Virology* 2003, 314:403-409.

[164] Shah KV, McCrumb FR, Jr., Daniel RW, Ozer HL: Serologic evidence for a simian-virus-40-like infection of man. *J Natl Cancer Inst* 1972, 48:557-561.

[165] Butel JS, Arrington AS, Wong C, Lednicky JA, Finegold MJ: Molecular evidence of simian virus 40 infections in children. *J Infect Dis* 1999, 180:884-887.

[166] Jafar S, Rodriguez-Barradas M, Graham DY, Butel JS: Serological evidence of SV40 infections in HIV-infected and HIV-negative adults. *J Med Virol* 1998, 54:276-284.

[167] Knowles WA, Pipkin P, Andrews N, Vyse A, Minor P, Brown DW, Miller E: Population-based study of antibody to the human polyomaviruses BKV and JCV and the simian polyomavirus SV40. *J Med Virol* 2003, 71:115-123.

[168] Minor P, Pipkin P, Jarzebek Z, Knowles W: Studies of neutralising antibodies to SV40 in human sera. *J Med Virol* 2003, 70:490-495.

[169] Carter JJ, Madeleine MM, Wipf GC, Garcea RL, Pipkin PA, Minor PD, Galloway DA: Lack of serologic evidence for prevalent simian virus 40 infection in humans. *J Natl Cancer Inst* 2003, 95:1522-1530.

[170] Stratton K, Almario D, McCormick M: *SV40 contamination of polio vaccine and cancer.* Washington, DC: The National Academies Press; 2002.

[171] Carbone M, Rizzo P, Procopio A, Giuliano M, Pass HI, Gebhardt MC, Mangham C, Hansen M, Malkin DF, Bushart G, et al: SV40-like sequences in human bone tumors. *Oncogene* 1996, 13:527-535.

[172] Meneses A, Lopez-Terrada D, Zanwar P, Killen DE, Monterroso V, Butel JS, Vilchez RA: Lymphoproliferative disorders in Costa Rica and simian virus 40. *Haematologica* 2005, 90:1635-1642.

[173] Vilchez RA, Butel JS: Simian virus 40 and its association with human lymphomas. *Curr Oncol Rep* 2003, 5:372-379.

[174] Vilchez RA, Lopez-Terrada D, Middleton JR, Finch CJ, Killen DE, Zanwar P, Jorgensen JL, Butel JS: Simian virus 40 tumor antigen expression and immunophenotypic profile of AIDS-related non-Hodgkin's lymphoma. *Virology* 2005, 342:38-46.

[175] Vilchez RA, Madden CR, Kozinetz CA, Halvorson SJ, White ZS, Jorgensen JL, Finch CJ, Butel JS: Association between simian virus 40 and non-Hodgkin lymphoma. *Lancet* 2002, 359:817-823.

[176] Carbone M, Bocchetta M, Cristaudo A, Emri S, Gazdar A, Jasani B, Lednicky J, Miele L, Mutti L, Pass HI, et al: SV40 and human brain tumors. *Int J Cancer* 2003, 106:140-142; author reply 143-145.

[177] Carbone M, Pass HI, Rizzo P, Marinetti M, Di Muzio M, Mew DJ, Levine AS, Procopio A: Simian virus 40-like DNA sequences in human pleural mesothelioma. *Oncogene* 1994, 9:1781-1790.

[178] Lednicky JA, Garcea RL, Bergsagel DJ, Butel JS: Natural simian virus 40 strains are present in human choroid plexus and ependymoma tumors. *Virology* 1995, 212:710-717.

[179] Bergsagel DJ, Finegold MJ, Butel JS, Kupsky WJ, Garcea RL: DNA sequences similar to those of simian virus 40 in ependymomas and choroid plexus tumors of childhood. *N Engl J Med* 1992, 326:988-993.

[180] Krieg P, Amtmann E, Jonas D, Fischer H, Zang K, Sauer G: Episomal simian virus 40 genomes in human brain tumors. *Proc Natl Acad Sci U S A* 1981, 78:6446-6450.

[181] Shivapurkar N, Harada K, Reddy J, Scheuermann RH, Xu Y, McKenna RW, Milchgrub S, Kroft SH, Feng Z, Gazdar AF: Presence of simian virus 40 DNA sequences in human lymphomas. *Lancet* 2002, 359:851-852.

[182] Shivapurkar N, Takahashi T, Reddy J, Zheng Y, Stastny V, Collins R, Toyooka S, Suzuki M, Parikh G, Asplund S, et al: Presence of simian virus 40 DNA sequences in human lymphoid and hematopoietic malignancies and their relationship to aberrant promoter methylation of multiple genes. *Cancer Res* 2004, 64:3757-3760.

[183] Shivapurkar N, Wiethege T, Wistuba, II, Salomon E, Milchgrub S, Muller KM, Churg A, Pass H, Gazdar AF: Presence of simian virus 40 sequences in malignant mesotheliomas and mesothelial cell proliferations. *J Cell Biochem* 1999, 76:181-188.

[184] Suzuki M, Toyooka S, Shivapurkar N, Shigematsu H, Miyajima K, Takahashi T, Stastny V, Zern AL, Fujisawa T, Pass HI, et al: Aberrant methylation profile of human malignant mesotheliomas and its relationship to SV40 infection. *Oncogene* 2005, 24:1302-1308.

[185] Toyooka S, Pass HI, Shivapurkar N, Fukuyama Y, Maruyama R, Toyooka KO, Gilcrease M, Farinas A, Minna JD, Gazdar AF: Aberrant methylation and simian virus 40 tag sequences in malignant mesothelioma. *Cancer Res* 2001, 61:5727-5730.

[186] Girardi AJ, Sweet BH, Slotnick VB, Hilleman MR: Development of tumors in hamsters inoculated in the neonatal period with vacuolating virus, SV-40. *Proc Soc Exp Biol Med* 1962, 109:649-660.

[187] Thu GO, Hem LY, Hansen S, Moller B, Norstein J, Nokleby H, Grotmol T: Is there an association between SV40 contaminated polio vaccine and lymphoproliferative disorders? An age-period-cohort analysis on Norwegian data from 1953 to 1997. *Int J Cancer* 2006, 118:2035-2039.

[188] Hachana M, Trimeche M, Ziadi S, Amara K, Korbi S: Evidence for a role of the Simian Virus 40 in human breast carcinomas. *Breast Cancer Res Treat* 2008.

[189] Giuliani L, Jaxmar T, Casadio C, Gariglio M, Manna A, D'Antonio D, Syrjanen K, Favalli C, Ciotti M: Detection of oncogenic viruses SV40, BKV, JCV, HCMV, HPV and p53 codon 72 polymorphism in lung carcinoma. *Lung Cancer* 2007, 57:273-281.

[190] Vilchez RA, Kozinetz CA, Arrington AS, Madden CR, Butel JS: Simian virus 40 in human cancers. *Am J Med* 2003, 114:675-684.

[191] Leithner K, Leithner A, Clar H, Weinhaeusel A, Radl R, Krippl P, Rehak P, Windhager R, Haas OA, Olschewski H: Mesothelioma mortality in Europe: impact of asbestos consumption and simian virus 40. *Orphanet J Rare Dis* 2006, 1:44.

[192] Kjaerheim K, Roe OD, Waterboer T, Sehr P, Rizk R, Dai HY, Sandeck H, Larsson E, Andersen A, Boffetta P, Pawlita M: Absence of SV40 antibodies or DNA fragments in prediagnostic mesothelioma serum samples. *Int J Cancer* 2007, 120:2459-2465.

[193] Price MJ, Darnton AJ, McElvenny DM, Hodgson JT: Simian virus 40 and mesothelioma in Great Britain. *Occup Med (Lond)* 2007, 57:564-568.

[194] Zekri AR, Bahnassy AA, Mohamed WS, Hassan N, Abdel-Rahman AR, El-Kassem FA, Gaafar R: Evaluation of simian virus-40 as a biological prognostic factor in Egyptian patients with malignant pleural mesothelioma. *Pathol Int* 2007, 57:493-501.

[195] Zekri AR, Mohamed W, Bahnassy A, Refat L, Khaled M, Shalaby S, Hafez M: Detection of simian virus 40 DNA sequences in Egyptian patients with different hematological malignancies. *Leuk Lymphoma* 2007, 48:1828-1834.

[196] Comar M, Rizzardi C, de Zotti R, Melato M, Bovenzi M, Butel JS, Campello C: SV40 multiple tissue infection and asbestos exposure in a hyperendemic area for malignant mesothelioma. *Cancer Res* 2007, 67:8456-8459.

[197] Cristaudo A, Foddis R, Vivaldi A, Buselli R, Gattini V, Guglielmi G, Cosentino F, Ottenga F, Ciancia E, Libener R, et al: SV40 enhances the risk of malignant mesothelioma among people exposed to asbestos: a molecular epidemiologic case-control study. *Cancer Res* 2005, 65:3049-3052.

[198] Brown P, Tsai T, Gajdusek DC: Seroepidemiology of human papovaviruses. Discovery of virgin populations and some unusual patterns of antibody prevalence among remote peoples of the world. *Am J Epidemiol* 1975, 102:331-340.

[199] Shah KV, Ozer HL, Ghazey HN, Kelly TJ, Jr.: Common structural antigen of papovaviruses of the simian virus 40-polyoma subgroup. *J Virol* 1977, 21:179-186.

[200] Hamilton RS, Gravell M, Major EO: Comparison of antibody titers determined by hemagglutination inhibition and enzyme immunoassay for JC virus and BK virus. *J Clin Microbiol* 2000, 38:105-109.

[201] Viscidi RP, Rollison DE, Viscidi E, Clayman B, Rubalcaba E, Daniel R, Major EO, Shah KV: Serological cross-reactivities between antibodies to simian virus 40, BK virus, and JC virus assessed by virus-like-particle-based enzyme immunoassays. *Clin Diagn Lab Immunol* 2003, 10:278-285.

[202] Lundstig A, Eliasson L, Lehtinen M, Sasnauskas K, Koskela P, Dillner J: Prevalence and stability of human serum antibodies to simian virus 40 VP1 virus-like particles. *J Gen Virol* 2005, 86:1703-1708.

[203] Rollison DE, Helzlsouer KJ, Alberg AJ, Hoffman S, Hou J, Daniel R, Shah KV, Major EO: Serum antibodies to JC virus, BK virus, simian virus 40, and the risk of incident adult astrocytic brain tumors. *Cancer Epidemiol Biomarkers Prev* 2003, 12:460-463.

[204] Rollison DE, Helzlsouer KJ, Halsey NA, Shah KV, Viscidi RP: Markers of past infection with simian virus 40 (SV40) and risk of incident non-Hodgkin lymphoma in a Maryland cohort. *Cancer Epidemiol Biomarkers Prev* 2005, 14:1448-1452.

[205] de Sanjose S, Shah KV, Domingo-Domenech E, Engels EA, Fernandez de Sevilla A, Alvaro T, Garcia-Villanueva M, Romagosa V, Gonzalez-Barca E, Viscidi RP: Lack of

serological evidence for an association between simian virus 40 and lymphoma. *Int J Cancer* 2003, 104:522-524.

[206] Engels EA, Chen J, Hartge P, Cerhan JR, Davis S, Severson RK, Cozen W, Viscidi RP: Antibody responses to simian virus 40 T antigen: a case-control study of non-Hodgkin lymphoma. *Cancer Epidemiol Biomarkers Prev* 2005, 14:521-524.

[207] Engels EA, Switzer WM, Heneine W, Viscidi RP: Serologic evidence for exposure to simian virus 40 in North American zoo workers. *J Infect Dis* 2004, 190:2065-2069.

[208] Engels EA, Viscidi RP, Galloway DA, Carter JJ, Cerhan JR, Davis S, Cozen W, Severson RK, de Sanjose S, Colt JS, Hartge P: Case-control study of simian virus 40 and non-Hodgkin lymphoma in the United States. *J Natl Cancer Inst* 2004, 96:1368-1374.

[209] Testa J, Carbone M, Hirvonen A, Khalili K, Krynska B, Linnainmaa K, Pooley F, Rizzo P, Rusch V, Xiao G: A multi-institutional study confirms the presence and expression of simian virus 40 in human malignant mesotheliomas. *Cancer Res* 1998, 58:4505-4509.

[210] Strickler HD: A multicenter evaluation of assays for detection of SV40 DNA and results in masked mesothelioma specimens. *Cancer Epidemiol Biomarkers Prev* 2001, 10:523-532.

[211] Lopez-Rios F, Illei PB, Rusch V, Ladanyi M: Evidence against a role for SV40 infection in human mesotheliomas and high risk of false-positive PCR results owing to presence of SV40 sequences in common laboratory plasmids. *Lancet* 2004, 364:1157-1166.

[212] Jasani B, Jones CJ, Radu C, Wynford-Thomas D, Navabi H, Mason M, Adams M, Gibbs A: Simian virus 40 detection in human mesothelioma: reliability and significance of the available molecular evidence. *Front Biosci* 2001, 6:E12-22.

[213] Shivapurkar N, Wiethege T, Wistuba I, Salomon E, Milchgrub S, Muller K, Churg A, Pass H, Gazdar A: Presence of simian virus 40 sequences in malignant mesotheliomas and mesothelial cell proliferations. *J Cell Biochem* 1999, 76:181-188.

[214] Cutrone R, Lednicky J, Dunn G, Rizzo P, Bocchetta M, Chumakov K, Minor P, Carbone M: Some oral poliovirus vaccines were contaminated with infectious SV40 after 1961. *Cancer Res* 2005, 65:10273-10279.

[215] Nurgalieva ZZ, Wong C, Zhangabylov AK, Omarbekova ZE, Graham DY, Vilchez RA, Butel JS: Polyomavirus SV40 infections in Kazakhstan. *J Infect* 2005, 50:142-148.

[216] Innis MD: Oncogenesis and poliomyelitis vaccine. *Nature* 1968, 219:972-973.

[217] Heinonen OP, Shapiro S, Monson RR, Hartz SC, Rosenberg L, Slone D: Immunization during pregnancy against poliomyelitis and influenza in relation to childhood malignancy. *Int J Epidemiol* 1973, 2:229-235.

In: DNA Tumor Viruses
Editor: H. E. Tao
ISBN 978-1-60692-111-1

Chapter II

NEW INSIGHTS INTO THE ROLE OF THE dUTPASE IN EPSTEIN-BARR VIRUS REPLICATION AND PATHOGENESIS

Marshall Williams[1,2,*] and Ronald Glaser[1,2,3]
[1]The Department of Molecular Virology, Immunology and Medical Genetics,
[2]The Comprehensive Cancer Center, The Ohio State University Medical Center and
[3]The Institute for Behavioral Medicine Research The Ohio State University Medical Center, USA.

ABSTRACT

Epstein-Barr virus (EBV), a gamma herpesvirus, is implicated in the pathogenesis of a variety of human malignancies including Burkitt's lymphoma (BL), nasopharyngeal carcinoma (NPC), Hodgkin's disease, and post-transplant B-cell lymphoma. Studies examining the mechanisms by which EBV causes malignancies have focused on those viral encoded proteins that are expressed during latency but there have been few studies to determine the potential role that proteins produced during lytic replication of EBV may have in the pathophysiology of EBV-associated diseases. We have demonstrated that the EBV-encoded deoxyuridine triphosphate nucleotidohydrolase (dUTPase), an enzyme that is produced during lytic replication of EBV, causes immune dysregulation in addition to its enzymatic function. The EBV-encoded dUTPase induces the upregulation of several proinflammatory cytokines including TNF-α, IL-1β, IL-6 and IL-8 as well as IL-10 in human monocytes/macrophages through the activation of NF-κB. Furthermore, the EBV-encoded dUTPase inhibited human T-cell proliferation in vitro. The EBV-encoded dUTPase also inhibited the replication of mitogen-stimulated lymphocytes and the synthesis of interferon-γ by cells isolated from lymph nodes and spleens of mice treated with the dUTPase. The data provide a new perspective on how a nonstructural

* Correspondence concerning this article should be addressed to: Marshall Williams, PhD, The Ohio State University, Department of Molecular Virology, Immunology and Medical Genetics, 2074 Graves Hall 333 W. 10th Ave, Columbus, OH 43210, Phone: (614) 292-0717; Fax: (614) 292-9805; e-mail: Williams.70@osu.edu.

protein that is associated with lytic replication of EBV can cause immune dysregulation. While the production of these proinflammatory cytokines may contribute to the pathology associated with EBV infection, perhaps more importantly, the production of IL-10 may alter the response of T-cells, ($CD4^+$ and $CD8^+$) to EBV infection. This may create a favorable environment for controlling the steady state expression of latent EBV and the ability of EBV genome positive tumor cells to survive.

INTRODUCTION

EBV is an oncogenic gamma herpesvirus that infects a significant percentage (>90%) of the worldwide population. EBV is similar to other herpesviruses in that following primary infection, the virus establishes where it can be reactivated during the lifetime of the individual. While EBV is considered to be a B-lymphotrophic virus, it is known to infect epithelial cells [Glaser et al, 1976; 1980; Bonnet et al., 1999; Walling et al, 2001] and there is increasing data demonstrating that EBV infects monocytes, dendritic cells, as well as T-cells *in vitro* [Nakamura et al, 1998; Gosselin et al, 1991b, 1992; Savard et al 2000, 2006; Masy et al, 2002; Lund et al, 2003; Yang et al, 2004].

In a primary infection the virus replicates initially in the oropharnyx probably in an epithelial cell that is either directly infected by the virus or through the transfer of the virus from the surface of a B cell [Shannon-Lowe et al, 2006]. Lytic replication results in the amplification of the input virus and shedding of the virus. EBV also infects mucosal B cells in the oropharnyx where it results in the polyclonal expansion of the lymphoidblastoid cells, which is the result of expression of the latency III program (growth program) [Keiff and Rickinson, 2001; Young and Rickinson, 2004]. While most of these cells are eliminated by the host's immune response, some cells survive because latent antigen expression of the virus is downregulated (Latency II). Ultimately, the expression of the EBV genome is further downregulated (latency I) and these cells enter the circulation as long-lived memory B cells. In healthy immunocompetent individuals the frequency of memory B cells latently infected with EBV ranges from 0.1 to 1×10^6 cells and it is approximately 50 times higher in immunosuppressed patients. EBV remains in this quiescent state until these memory B cells are activated and terminally differentiated into plasma cells and then the virus undergoes lytic replication [Thorley-Lawson, 2001; Pegtel et al, 2004; Laichalk and Thorley-Lawson, 2005]. This cycle results in maintenance of the persistent infection and allows period shedding of the virus in saliva.

Primary infection with EBV usually occurs within the first few years of life in under-developed countries and is generally asymptomatic. In developed countries primary infection with EBV tends to occur later, late adolescence or adulthood. In most cases the infection is asymptomatic but infectious mononucleosis (IM) occurs in some cases. In addition to IM EBV is implicated in the pathogenesis of post-transplant lymphoproliferative disorders, as well as, a variety of human malignancies including Burkitt's lymphoma (BL), nasopharyngeal carcinoma (NPC), Hodgkin's disease, chronic lymphocytic leukemia (CLL), non-Hodgkin's lymphoma and gastric cancer [Ablashi et al., 1990; Petrella et al., 1997; Ansell et al., 1999; Bonnet et al., 1999; Brousset et al., 2002; Iwakiri et al., 2003; Thompson and Kurzrock, 2004]. In recent years EBV-genome-positive malignancies have become increasingly common especially in transplant patients and patients with acquired immunodeficiency syndrome [Ho et al., 1998; Touitou et al., 2003].

Studies to examine the roles of various proteins in cellular transformation have focused primarily on those proteins and RNAs expressed during latency. These studies have demonstrated unequivocally the role that latent membrane protein –1 (LMP1) and LMP-2A have in the transformation process and the immunological response of the host to these proteins [Khanna and Burrows, 2000; Hislop et al., 2007]. Conversely, there have been very few studies directed toward determining the role(s), in immune modulation or in transformation of proteins expressed during the lytic replication. This is due in part to the lack of model systems to examine lytic replication of EBV in vitro. However, recent studies are demonstrating that EBV-encoded proteins expressed during lytic replication may contribute to the pathophysiology observed in infections with EBV.

Lytic Replication of EBV

The establishment and maintenance of persistent EBV infections requires lytic replication of the virus. In the immunocompetent healthy individual this occurs primarily in the oropharyngeal epithelium with shedding of virus in the salvia [Thorley-Lawson, 2001; Pegtel et al, 2004; Laichalk and Thorley-Lawson, 2005]. Lytic replication of EBV has also been reported to occur in the epithelial cells of patients with hairy leukoplakia (HL), a proliferative disease of the squamous epithelial cells of the lateral tongue [Webster-Cyriaque et al, 2000] and in epithelial cells of patients with breast cancer [Bonnet, et al 1999; Huang et al, 2003]. In previous work from our lab and others, it was demonstrated that psychological stress could reactivate latent EBV and evidence for abortive reactivation was found [Esterling et al., 1990; 1992; Glaser et al., 1993; 1994; 2001; 2005; Glaser and Kiecolt-Glaser, 2005; Stowe et al., 2007]. Likewise, it was recently reported that the presence of lytic replication-defective human herpesvirus -8 (HHV-8), another human oncogenic gamma-herpesvirus, in Kaposi's sarcoma tumors and primary effusion lymphoma (PEL) cell lines enhanced cell proliferation and transforming potential [Blasig et al, 1997; Deng et al, 2004]. It seems logical that replication-defective mutants of EBV would be generated over the course of a lifetime in latently infected individuals and that while progeny viruses would not be generated with the induction of lytic replication, the uncontrolled expression of genes encoding for immediately early and early proteins involved with DNA replication could contribute to the pathology of EBV associated diseases.

Antibody titers to various EBV antigens have been used to diagnose a variety of diseases for which EBV has been associated. Commonly observed in patients acutely or latently infected with EBV are antibodies to several early antigens (EA), as well as antibodies to virus capsid antigen (VCA) structural proteins. In addition, antibodies to the family of nuclear antigens expressed in cells latently infected with EBV are also observed in individuals latently infected with EBV; these antibodies are called EBNA antibodies [Reedman and Klein, 1973]. Of interest is the fact that patients with African BL and patients with EBV-associated NPC have unique antibody patterns to the EBV EA complex. Patients with BL tend to make antibody to the restricted (R) component of the EA complex (along with VCA) and patients with NPC tend to have antibody titers to the diffuse (D) component of the EA complex along with VCA [Henle et al, 1971; 1973]. These antibody patterns have been used diagnostically [Henle et al, 1971; 1973; Levine et al, 1985]. A recent multicentric case-control study has demonstrated abnormal antibody patterns to components in EA in patients with a variety of lymphomas, suggesting that EBV might be involved in a larger proportion of lymphomas in clinically immunocompetent patients [deSanjose et al, 2007]. The reason for these antibody patterns in such patients and how they relate to virus replication or to the pathophysiology that occurs in EBV related diseases is unknown.

EBV encodes for several viral enzymes that are part of the EA complex. These include thymidine kinase (TK), DNA polymerase, deoxyribonuclease (DNase), deoxyuridine triphosphate nucleotidohydrolase (dUTPase), uracil-DNA glycosylse and ribonucleotide reductase [Glaser et al., 1973a, b; Miller et al., 1977; Henry et al., 1978; Clough et al., 1979; Cheng et al., 1980a; Williams et al., 1985a; Lu et al., 2007]. While the synthesis of EBV-encoded enzymes takes place in the early phase of the lytic replicative cycle of the virus, by definition, the EA complex of proteins do not require the replication of viral DNA prior to expression of these proteins, while the VCA complex of late proteins require the replication of EBV DNA, which encodes for those proteins.

Antibodies to several EBV-encoded enzymes have been observed in patients with different EBV-associated diseases. For example, antibodies to EBV DNA polymerase and DNase are observed in patients with NPC [Cheng et al., 1980b; Liu et al., 1989] and in patients with chronic fatigue syndrome [Jones et al., 1988]. There is a report that shows that NPC patients may also have antibody to EBV TK [Turenne-Tessier et al., 1989]. Patients with IM, chronic EBV infection, and patients infected with HIV have elevated antibody titers to EBV dUTPase [Sommer et al, 1996; Nicholls et al, 1998; Fleischmann et al, 2002] While these unique antibody patterns to different EBV proteins have been found clinically useful [Cheng et al., 1980b; Paramita et al, 2007], the underlying factors that result in these antibody patterns to EBV enzymes and the role these proteins might play, separate from their role in the replication of the virus, has not been explored in the pathophysiology of EBV-associated disease.

dUTPases

With the exception of some members of the genera *Mollicutes* [Williams and Pollack, 1985], all free-living organisms, as well as, some viruses encode for a dUTPase [Nyman, 2001]. dUTPases represent a family of metalloenzymes that catalyze the hydrolysis of dUTP to dUMP and pyrophosphate [Nyman, 2001] thus preventing dUTP from being incorporated into DNA by DNA polymerases. dUTPases can be divided into at least three subgroups based upon their structure and specificity for dUTP. The largest subgroup, which exhibits a high specificity for dUTP, is the homotrimeric dUTPases, which are found in plants, animals, fungi, bacteria and some viruses, including the retroviruses and poxviruses. These homotrimeric dUTPases are composed of three identical polypeptides aligned so that three highly conserved amino acids domains (domains 1, 2 and 4) on one polypeptide interact with the amino acids in conserved domain 3 on an adjacent peptide. The catalytic site is completed with the binding of dUTP to the flexible C-terminal end of the third polypeptide, which contains the conserved amino acids in domain 5. Thus, all three polypeptide subunits contribute to the formation of a catalytic site and there are three catalytic sites per holoenzyme. The homotrimeric dUTPase is essential for the replication of *E. coli* [El-Hajj et al, 1988] and *Saccharomyces cerevisiae* [Gadsden et al, 1993] and probably for human cells since no mutants have been isolated deficient in this activity. The monomeric dUTPases, which are thought to have arisen from the homotrimeric dUTPases by gene duplication [Baldo and McClure, 1999], are found exclusively in herpesviruses [McGeoch et al, 1995]. Structural data of the EBV-encoded dUTPase, which is the smallest of the herpesviruses-encoded dUTPases, has demonstrated that the single catalytic site of the enzyme is comprised of the five highly conserved domains characteristic of the homotrimeric and monomeric dUTPase and thus the catalytic site of EBV-encoded mimics the catalytic site of the homotrimeric dUTPases [Tarbouirech et al, 2005]. Sequencing analyses have recently demonstrated that the herpesviruses' dUTPases contain an additional conserved domain (domain 6) that is absent in the homotrimeric dUTPases [Davidson and Stowe, 2005]. It has been suggested that this novel herpesvirus-specific domain may contribute to some novel function but that function(s) has not been elucidated. The last group, the homodimeric dUTPases, identified in *Leshmania major* [Comacho et al, 1997; 2000], *Trypansoma cruzi* [Bernier-Villamor et al, 2002], *Caenorhabditis elegans* [Parkhill et al, 2000] and recently in *Campylobacter jejuni* [Hill et al., 1998] differ not only structurally from the monomeric and homotrimeric dUTPases, but they also exhibit broader substrate specificity. Furthermore, sequence comparisons have demonstrated that the homodimeric dUTPases lack the five consensus amino acid motifs found in mono and trimeric dUTPases and that they are evolutionary related to the dCTPase-dUTPase of bacteriophages T2 and T4 [Comacho et al, 2000].

The role(s) that virus-encoded dUTPases have in virus replication processes and in pathogenesis is unclear. It is generally assumed that the primary function of these virus-encoded dUTPases, including the EBV-encoded dUTPase, is to maintain a low dUTP pool during virus replication, thus preventing dUTP incorporation into the replicating viral genome. However, the critical studies to demonstrate this premise, as well as, studies to examine alterations in dUTP metabolism following virus infections have not been performed. In the case of the EBV-encoded dUTPase, it is established that the EBV-encoded dUTPase is expressed *in vivo* [Sommer et al, 1996; Nicholls et al, 1998; Fleischmann et al, 2002] and our recent studies have demonstrated the EBV-encoded dUTPase not only modulates the host's immune system, but its expression may also contribute the pathology of EBV associated diseases [Padgett et al., 2004; Glaser et al., 2006; Waldman et al., 2007]

EBV-Encoded dUTPase Modulates Innate Immunity

The innate immune response is an early line of defense, which plays a key role in the protection of a host from invading pathogens including viruses [Kawai and Akira, 2006; Fineberg et al, 2007]. The innate immune response to viruses is generally characterized by the production of proinflammatory chemokines and cytokines as well as by the production of type 1 interferons (interferon α and β). However, the host response to viruses is dependent not only upon the cell type involved and also upon whether the virus has mechanisms to counter the response. Viruses, as well as other pathogens, encode for various proteins containing pathogen associated molecular patterns (PAMPs), which are recognized by immune–sensor molecules that are referred to as pattern recognition receptors (PRRs). Toll-like receptors (TLR), which are PRRs, are responsible for the primary recognition of a broad range of pathogens [Akira and Takedo, 2004; Finberg et al, 2007], leading to the initiation of the innate and adaptive immune response [Akira and Takedo, 2004; Beutler, 2004; Finberg et al, 2007. When TLRs are triggered signaling cascades are activated leading to the enhanced transcription of proinflammatory cytokines genes or the production of type 1 interferons. While this is usually a protective mechanism in the case of some herpesviruses, a TLR mediated cytokine response, contributes to neuropathogenesis in the host [Aravalli et al., 2005; Kurt-Jones et al., 2005; Mansur et al., 2005]

There are numerous reports showing that acute and chronic EBV infections result in changes in the secretion patterns of TNF-α, IL-1β, IL-6, and IL-10 [Gosselin et al., 1991; 1992; Sharma and Zheng, 2001; Mori et al., 2003; Lucian et al, 1998; Klein et al, 1992; Fayad et al, 2001; Lai et al, 2002; Miyazaki et al, 1993; Budiani et al, 2002]. As discussed previously EBV is associated with several malignant diseases including NPC, chronic lymphocytic leukemia (CLL), and several B-cell malignancies including Burkitt's lymphoma (BL) and post-transplant B-cell lymphomas [Ablashi et al, 1990; Petrella et al, 1997; Ansell et al., 1999 Brousset et al, 2002; Iwakiri et al, 2003; Thompson and Kurzrock, 2004;] There is evidence linking serum IL-6 and IL-10 levels with prognosis in some of these cancers. Serum levels for both IL-6 and IL-10 have been correlated with disease outcome in CLL and Hodgkin's disease patients with higher levels predictive of a poorer health outcome [Fayad et al., 2001; Lai et al., 2002]. Serum from patients with NPC have elevated levels of IL-10 that are related to late-stage disease, which suggests that IL-10 may have a potential role in the progression of NPC tumors [Budiani et al., 2002] and tumor biopsies showed that NPC tumor cells were positive for IL-10 [Budiani et al, 2002]. Furthermore, anti-IL-10 antibodies were detected in patients with chronic active EBV infections [Tanner et al., 1997]

There may also be a link for IL-10 in the pathogenesis of EBV-associated B-cell lymphomas [Khatri and Caliguiri, 1998]. Studies in mice persistently infected with lymphocytic choriomeningitis virus (LCMV) demonstrated that the establishment of the chronic infection was associated with the functional impairment and deletion of virus specific-$CD8^+$ T cells [Moskophidis et al., 1993; Zajac et al., 1998]. Recently it was shown that there was an increased expression of PD-1, an inhibitory receptor of the CD28 family in T cells in mice chronically infected with LCMV and that inhibition of the interaction of PD-1 with its ligand PD-L1 restored T cell function [Barber et al, 2006]. Other studies have subsequently demonstrated that IL-10 is upregulated in mice chronically infected with LCMV and that blockage of the IL-10 receptor results in restoration of T cell function [Brooks et al., 2006; Ejrnaes et al., 2006]. While additional studies are needed these results suggest that IL-10 and PD-1 expression are linked and maybe responsible for initiation and maintenance of viral persistence by causing CD8+ T cell dysfunction.

There have been numerous studies to identify the cells and the mechanism(s) by which EBV alters cytokine production since accumulating evidence suggests that cytokines may play an important role in the pathogenesis of EBV infections. Several studies have described demonstrated that EBV infects neutrophils resulting in the production of reactive oxygen species, the enhanced biosynthesis of leukotrienes and the increased synthesis of IL-8 and macrophage inflammatory protein-1α [Beaulieu et al., 1995; McColl et al., 1997; Gosselin and Borgeat, 1997; Gosselin et al., 2001; Savard and Gosselin, 2006]. Neutrophils infected by EBV rapidly undergo apoptosis preventing the completion of the viral replicative cycle and thus leading to an abortive infection [Larochelle et al., 1998]. This could release immediate early and early proteins encoded by EBV into the microenvironment and these EBV encoded proteins could possibly function as PAMPs thus amplifying the proinflammatory response to EBV. A similar mechanism has been used to explain how normal cellular endogenous proteins such as heat shock proteins 60 and 70 activate TLRs [Vabulas et al., 2001; Asea et al., 2002]. EBV productively infects monocytes and macrophages and following these infection there are changes in the expression of various cytokines/chemokines [Shimakage et al., 1999; D'Addario et al., 1999; 2000; 2001; Savard et al., 2000; 2006; Salek-Ardakani et.al, 2004; Gaudreault et al., 2007]. Recently, Gaudreault et al [2007] reported that EBV virions were recognized by TLR2 and that this interaction was blocked by the 72A1 antibody, which recognizes an epitope of gp350, suggesting that gp350 is acting like a PAMP. Martin et al. [2007] also reported that EBV up-regulated the expression of TLR7, down-regulated TLR9 expression in naïve B cells and that EBV modulates TLR7 signaling to enhance B cell proliferation and to regulate IRF-5 activity. Overall, these results suggest that there is a virion component, possibly gp350 that plays an important role(s) in the recognition of EBV by macrophages/B-cells.

It has been suggested that EBV induces the production of cytokines/chemokines in monocytes/macrophages by at two distinct mechanisms [Savard and Gosselin, 2006; Gaudreault et al, 2007]: the first is through the interaction of EBV envelope glycoproteins, such as gp350, with cellular receptors and a second mechanism that requires virus replication. However, the viral protein(s) responsible for eliciting the increased production of cytokines/chemokines remains unknown. There are recent studies, which suggest that other EBV-encoded proteins can alter immune function when tested *in vitro*. Data has been reported showing that purified EBV latent membrane protein-1 (LMP-1), which is expressed in latently infected cells, was able to suppress T-cell and natural killer (NK) cell responses [Dukers et al., 2000], induces IL-10, [Anatomy et al., 1994] IL-6 and IL-8 [Eliopoulous et al., 1999]. EBNA2, which is expressed during latency III, was reported to induce Type 1 interferons [Kanda et al., 1999]. BZLF1 protein, which initiates viral early gene expression modulates TNF-α signaling [Morrison et al., 2004] and negatively regulates type 1 interferon production [Hahn et al., 2005]. Since EBV encodes for approximately 70 polypeptides, undoubtedly, more than these will have the potential to modulate immune function.

To determine whether the EBV-encoded dUTPase could alter cytokine/chemokine expression, non-stimulated human peripheral blood mononuclear cells (PBMCs) were treated with purified EBV-encoded dUTPase and the production/expression of various cytokines were determined by ELISA. Treatment of PBMCs with the EBV-encoded dUTPase increased the expression/production of the proinflammatory cytokines TNF-α, IL-1β, IL-6 and IL-8, as well as, IL-10 (Figure 1). The production /release of TNF- α peaked at 24 h and quickly declined over time (Figure 1A). Likewise, as shown in Figures 1B-1E, EBV-encoded dUTPase induced IL-1β, IL-6, IL-8 and IL-10 production by human PBMCs, which reached a maximum at 24 hours and slowly declined over time. The production of IL-1β, IL-6, IL-8, and IL-10 proteins was confirmed by demonstrating an increase in mRNA levels (see figure insert). No IL-2, IL-4, IL-5 or IL-12 p70 was detected in cell supernatants treated with the EBV dUTPase. Pretreatment of the EBV dUTPase with the mAb, 7D6, which is specific for the dUTPase, inhibited TNF-α, IL-10, IL-1β, and IL-8 production by approximately 50 % (Table 1).

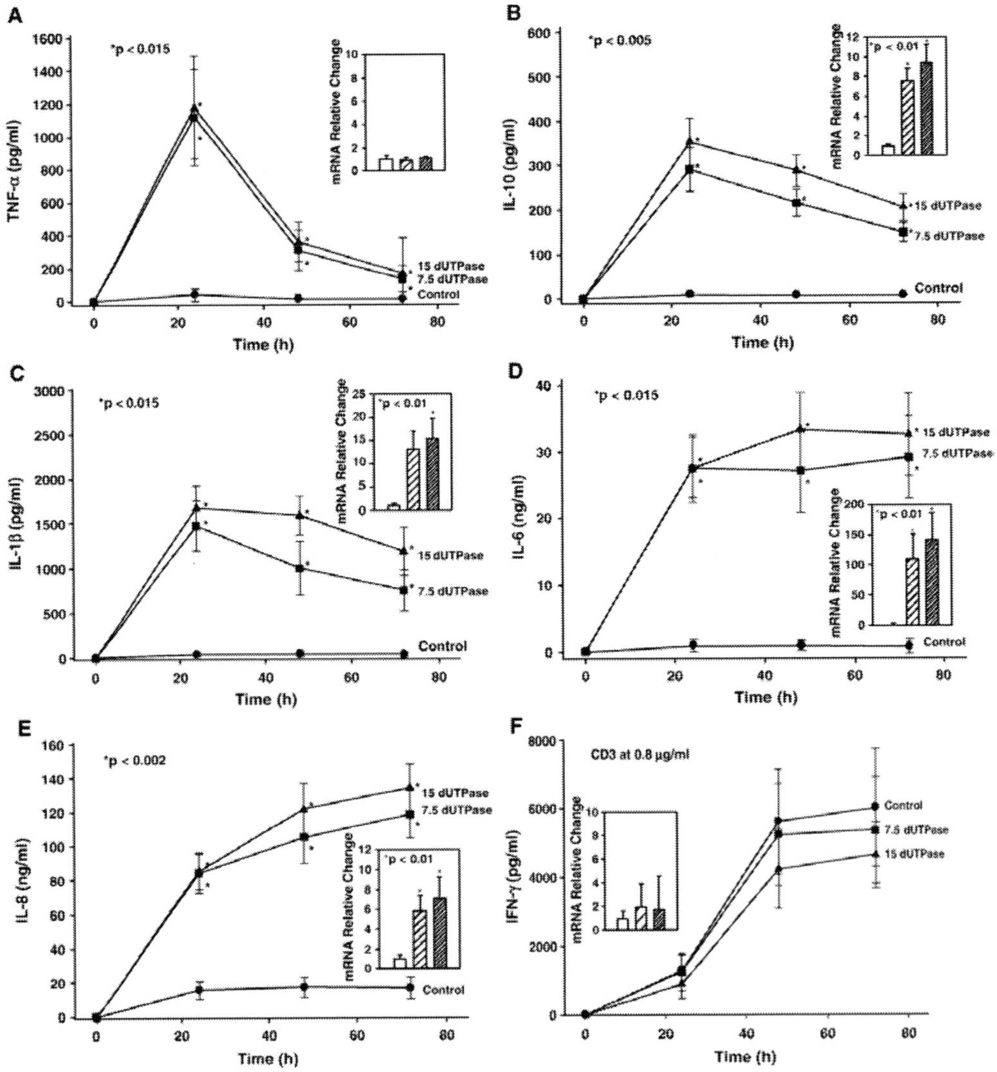

Figure 1A-F. The effect of EBV dUTPase on cytokine production. PBMCs were treated with two different concentrations of EBV dUTPase as described in the Methods. Untreated PBMCs (●), 7.5 μg/ml EBV dUTPase (■) and 15 μg/ml EBV dUTPase (▲). Cell supernatants were assayed for cytokine levels at 24, 48 and 72 hours. The results are the mean ± SEM; n=8. The inserted bar-graph plots represent the relative change in cytokine mRNA expression at 24 hours. *Untreated PBMCs (▭), 7.5 μg/ml EBV dUTPase (▨), and 15 μg/ml EBV dUTPase (▩).*

While the EBV-encoded dUTPase treatment of resting PBMCs induced proinflammatory cytokines, interferon-gamma (INF-γ) levels in dUTPase – treated cells were not significantly different from the control cultures (Figure 1F). Upon treatment of the cells with CD23 to induce T-cell blastogenesis, INF-γ levels increased 5- to 10- fold in control PBMCs in control cultures (six of eight individuals). Simultaneous treatment of the PBMCs with CD3 mAb and EBV encoded dUTPase resulted in lower INF-γ levels when compared to the control cultures from the same individual. While the results were not statistically different ($p < 0.084$ for 7.5 μg/ml and $p < 0.064$ for 15 μg/ml; $n = 8$), there was a trend showing that INF-γ production was suppressed after 48 h when the PBMCs were treated with the dUTPase.

Table 1. Antibody Neutralization of EBV dUTPase–induced Cytokine Profile

Antibody	Inhibition of EBV dUTPase-Induced Cytokine Production*				
	TNF-α	IL-10	IL-1β	IL-6	IL-8
Rat IgG1	0%	0%	4.2 ± 2.1%	4.3 ± 0.4%	3.7 ± 1.2%
Monoclonal anti-EBVdUTPase 7D6	59.2 ± 13.5%	46.4 ± 3.8%	59.6 ± 2.3%	21.9 ± 3.7%	45.7 ± 10.2%

*EBV dUTPase treatment was at 10 μg/ml for 24 hours. Data shown as the average ± SEM, n = 3.

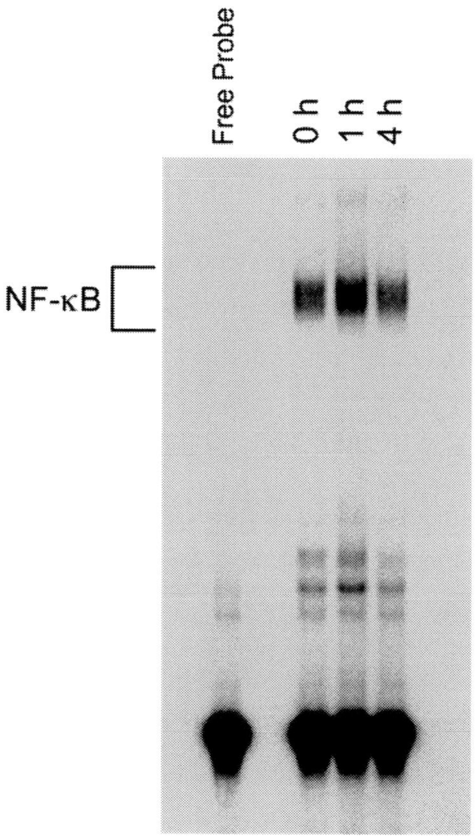

Figure 2. NF-κB is activated by dUTPase treatment of macrophages. Cells were treated with EBV-encoded dUTPase and, at indicated times, nuclear extracts were prepared and EMSA analysis was performed.

Since NF-κB is known to be a major transcriptional regulator in a number of inflammatory responses, in particular the activation of TNF-α and IL-6, we determined the effect of the EBV-encoded dUTPase in activating NF-κB. Treatment of PBMCs the EBV-encoded dUTPase resulted in the induction of NF-κB. Maximum induction occurred 1 after EBV-encoded dUTPase treatment, and which to basal levels after 4 hours of stimulation (Figure 2). Furthermore, EMSA demonstrated that activation of NF-κB comprised primarily of p50/c-Rel heterodimers with a slight contribution from p50/p65 complexes (Figure 3).

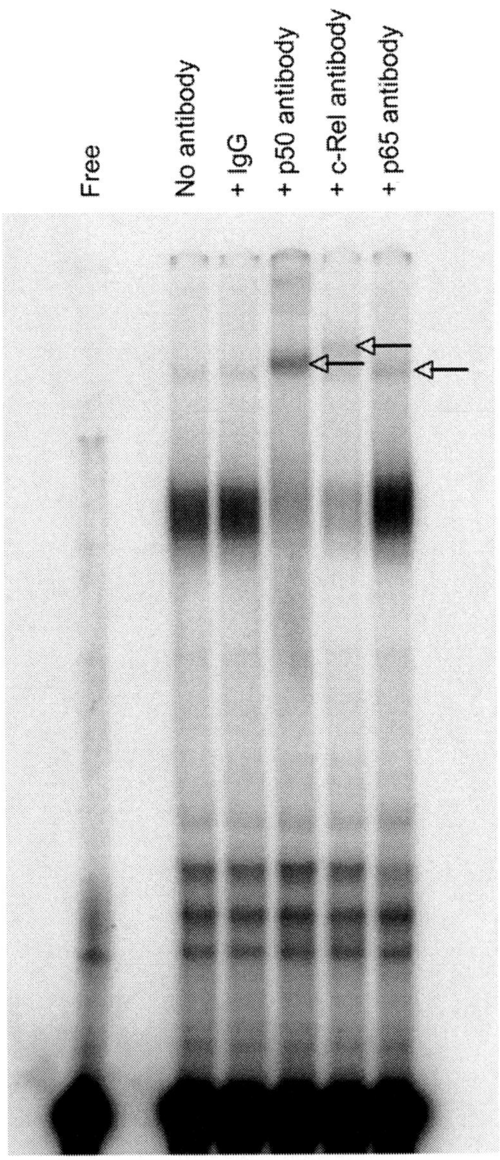

Figure 3. dUTPase activates the p50/c-Rel complex of NF-κB. Nuclear extracts were pre-incubated with antibodies specific to NF-κB subunits and EMSA analysis was subsequently performed. Arrows indicate supershifted complexes.

To confirm that NF-κB contributed to cytokine production, dUTPase-treated PBMCs were incubated in the presence of NF-κB inhibitors, PS1145 and NBD that target the IKK complex responsible for regulating NF-κB activity. In the presence of the inhibitors, TNF-α production was markedly reduced ($p < 0.001$), which corresponded to a concomitant reduction of NF-κB DNA binding activity ($p < 0.001$), and similar reductions were also observed for IL-6 (Figure 4 and data not shown). Together, these data demonstrate that the expression of proinflammatory cytokines is mediated through EBV-encoded dUTPase-induced activation of NF-κB.

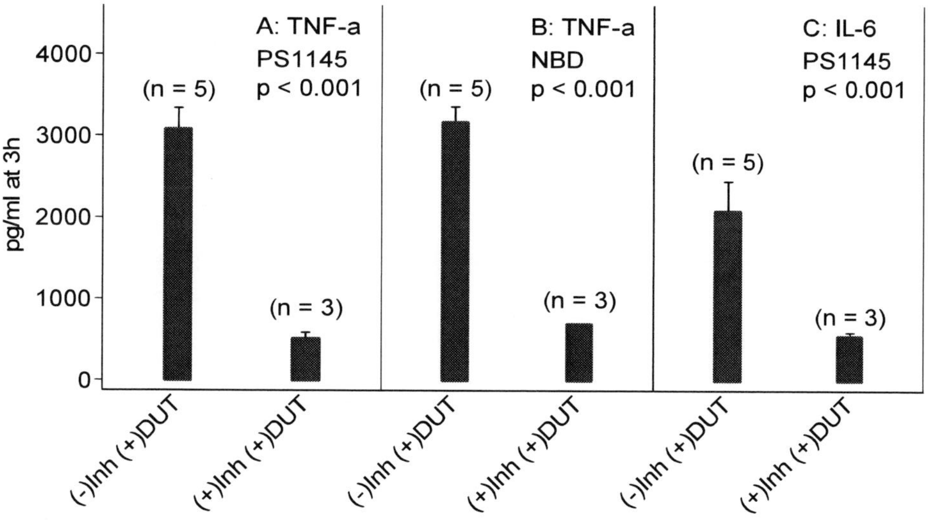

Figure 4. dUTPase-induced secretion of TNF-α and IL-6 from PBMCs is NF-κB dependent. TNF-α (A, B) and IL-6 (C) released from dUTPase-treated PBMCs pre-incubated with and without NF-κB inhibitor compounds (Lynch, 1972), EBV dUTPase treatment at 10 μg/ml, PS1145 (10 μM) and NBD (100 μM) . Inh=inhibitor, DUT=dUTPase.

The cytokine profile induced by treatment of PBMCs with the EBV-encoded dUTPase suggests involvement of monocytes/mcrophages. To test the hypothesis pure populations of CD20-positive cells (B cells) and CD14-positive cells (monocytes/macrophages) were obtained from freshly isolated PBMCs using Magnetic Cell Sorting (MACS) in the presence of FcR Blocking Reagent and either CD20 Microbeads or CD14 Microbeads Purity of the populations was confirmed by flow cytometry. The results shown in Table 2 are from one individual's PBMCs, and are representative of the results obtained with three different individuals. Monocytes/macrophages are the primary cell type affected by the EBV-encoded dUTPase. B-cells were also affected by treatment with the EBV-encoded dUTPase, but to a lesser extent and there was no induction of IL-10. However, the purified B-cells were assayed at a level that is ten times that within the mixed PBMC population. Finally the treatment of PBMCs that were depleted of both B-cells and monocytes/macrophages with the EBV-encoded dUTPase resulted in the increased production of TNF-α, IL-6 and IL-8. These results suggest that a third cell-type, which is adherent, is affected by the EBV-encoded dUTPase. Dendritic cells are the likely candidate based on adherence and antigen processing capabilities, and further studies are underway to purify the dendritic cells.

Table 2. Effect of EBV-encoded dUTPase on cytokine secretion by select cell populations[a]

		Cytokines at 48h (pg/ml)			
Cells (0.5 million/ml)	TNF-α	IL-10	IL-6	IL-1β	IL-8
Undepleted	814.0	515.0	14,433.0	2,093.0	86,840.0
CD20 positive	85.1	0.0	674.1	86.9	1,669.6
CD14 positive	1,510.0	165.0	25,670.0	1,365.0	176,500.0
CD14 and CD20 depleted	128.3	37.0	929.7	72.3	3,676.0

[a]Pure populations of CD20-positive cells (B-cells) and CD14-positive cells (monocytes/macrophages) were obtained using MACS in the presence of FcR Blocking Reagent in the presence of either CD20 Microbeads or CD14 Microbeads (Miltenyi Biotec, Germany) from freshly isolated PBMCs. Purity of the populations was confirmed by flow cytometry. The results are from one individual's PBMCs, and are representative of the results obtained with three different individuals.

These data also suggested that the EBV-encoded dUTPase was possibly activating NF-κB by triggering a TLR. To begin to address this possibility we treated human embryonic kidney cells (293) cells expressing various TLRs and adapter proteins and which were co-tranfected with a NF-κB-luciferase reporter plasmid with the EBV-encoded dUTPase examined the cells for the activation of NF-κB. While these studies are in progress the preliminary results, which are show in Table 3, suggest that the EBV-encoded dUTPase maybe activating NF-κB by triggering TLR2.

Table 3. Activation of NF-κB recorder gene in HEK293 cells expressing various TLRs and/or adaptor molecules

TLR/Adaptor Molecule	Relative NF-κB activity
none	1
TLR3	1
TLR4	1
TLR4/MD	1
TLR2	90
TLR2/CD14	90
TLR2-TLR1	90
TLR2-TLR6	25

The EBV-Encoded dUTPase and Cardiovascular Disease

Inflammation plays a central role in driving the evolution of atherosclerosis and coronary artery disease [Binder et al., 2002; Danesh et al., 1997; Rattazzi et al., 2005; Ridker et al., 2005a]. Clinical studies indicate that inflammation is an independent risk factor for the development of atherosclerosis and also interacts with established risk factors such as hypercholesterolemia [Binder et al., 2002; Ridker et al., 2005b; Steinberg, 2002]. Evidence from model systems and patient studies have demonstrated increases in the concentration and activity of factors that drive inflammation and innate immunity, including the well described increase in C-reactive protein (CRP), as well as upstream factors, such as IL-1β, IL-6, and TNF-α [Binder et al., 2002; Danesh et al., 1997; Kurzock, 2001; Rattazzi et al., 2005; Ridker et al., 2005a]. These cytokines are increased in circulating plasma in patients with cardiovascular disease and are present within atherosclerotic lesions themselves [Rattazzi et al., 2005]. In particular, TNF-α promotes the expression of pattern recognition receptors such as scavenger receptors and Toll-like receptors that facilitate the interaction between monocytes/macrophages and circulating lipid, thus leading to the formation of early atherosclerotic lesions [Rattazzi et al., 2005].

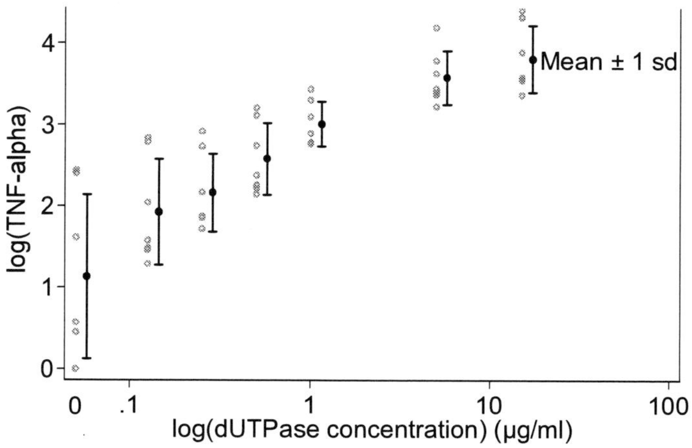

Figure 5A. EBV-encoded dUTPase stimulates TNF-α production by human differentiated macrophages. Human macrophages isolated and differentiated from 7 individual donors were treated for 24 hours with various concentrations of EBV-encoded dUTPase. TNF-α content in culture supernatants was measured by flow cytometry using the BD™ Human Inflammation Cytometric Bead Array. Open symbols represent log TNF-α values for each individual donor. Closed symbols represent mean log TNF-α values among donors at each dUTPase concentration ± 1 standard deviation.

The factors that promote the activation of these inflammatory mechanisms are incompletely defined. Viruses, such as EBV or cytomegalovirus (CMV), are a likely stimulus to inflammation, considering their persistence in the host and prevalence throughout the population; however, investigations have not provided consistent evidence that they play a significant role in atherogenesis [Danesh and Appleby, 1998; Horvath et al., 2000, Ibrahim et al., 2005]. A recent study demonstrating the presence of EBV DNA in atherosclerosis plaques is consistent with the interpretation that EBV can lytically replicate in the monocytes/macrophages [Savard et al., 2000], which have become associated with the plaque [deBoer et al., 2006]. EBV induces macrophages to synthesize the macrophage inflammatory protein-1 alpha (MIP-1α), which can attract B- and T-lymphocytes to the site of inflammation [McColl et al., 1997]; a recent study reported that EBV-infected B-cells could provide a source of EBV to infect monocytes [Tugizov et al., 2007]. Furthermore, the deBoer et al study showed that EBV-specific T-cells are generated at the site of the plaque as well, confirming that a local EBV-specific T-cell response can contribute to the inflammatory process presumably related to the EBV-infected macrophages [deBoer et al., 2006]. This investigation provides a plausible basis for the initiation of pro-atherogenic inflammatory responses by latent viral infections, such as EBV.

To begin to address the question of whether the expression of the EBV-encoded dUTPase could play a role in atherogenesis, monocyte derived mcrophages (MDM) were isolated and treated with various concentrations of the EBV-encoded dUTPase. As shown in the data (Figure 5) there was a dose-dependent increase in TNF-α production and an increase in IL-6 production in MDMs treated with the EBV-encoded dUTPase when compared to untreated controls.

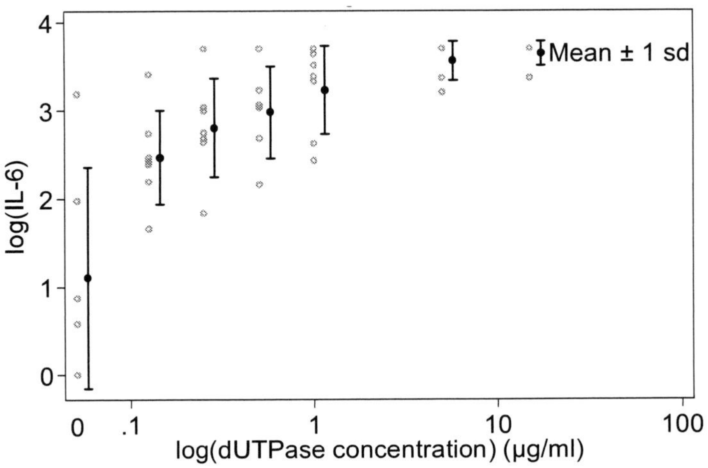

Figure 5B. EBV-encoded dUTPase stimulates IL-6 production by human differentiated macrophages. The conditions were the same as described in Figure 1A.

To simulate the impact of soluble factors produced by EBV-encoded dUTPase-exposed-MDMs upon proximal vascular endothelium, human umbilical vein endothelial cells (HUVEC) were incubated with supernatants of MDM treated for 24 hours with various concentrations of dUTPase (0.125 – 15 μg/ml). Controls included untreated HUVEC, HUVEC treated with 15 μg/ml dUTPase, and HUVEC treated with 300 IU/ml human recombinant TNF-α as a positive control for adhesion molecule induction. Following 8 (VCAM-1) or 24 (ICAM-1) hours of incubation, HUVEC were harvested and assayed forVCAM-1 or ICAM-1 expression by immunofluorescence flow cytometry. These incubation times were based upon numerous studies, which demonstrate that endothelial cells treated with TNF-α express peak levels of VCAM-1 at 8 hours and of ICAM-1 at 24 hours [Pober and Cotran, 1990]. Data generated by these experiments are presented in Figure 6 as the mean ± 1 standard deviation (solid symbols and error bars) for log endothelial ICAM-1 (6A) or VCAM-1 (6B) mean fluorescence intensity among the 7 MDM donors at each concentration of dUTPase. The magnitude of endothelial ICAM-1 and VCAM-1 induction by supernatants of EBV-encoded dUTPase-treated MDM varied among individual MDM donors, but exhibited a dose-response, which plateaus at the higher levels of dUTPase. The association between each of these markers and log(dUTPase) is quadratic (in all cases p <

0.0011), and in all cases the p-value for the repeated measures analysis of variance is significant at the $p < 0.0001$ level. For ICAM-1, the baseline level (0.05) differs significantly from all but the 0.125 and 0.25 levels of dUTPase. Log(ICAM-1) levels at the highest concentrations of dUTPase does not differ significantly from either of the next two highest concentrations suggesting that we had reached saturation. For VCAM-1, the baseline level (0.05) differs significantly from all higher levels of dUTPase. Log(VCAM-1) levels at the four highest concentrations of dUTPase do not differ significantly from one another again suggesting that we had reached saturation. The failure of direct dUTPase treatment of HUVEC to induce ICAM-1 or VCAM-1 verifies that endothelial adhesion molecules were induced by soluble factors produced by EBV-encoded dUTPase-treated MDM and not directly by residual dUTPase in MDM culture supernatants (data not shown).

A

B

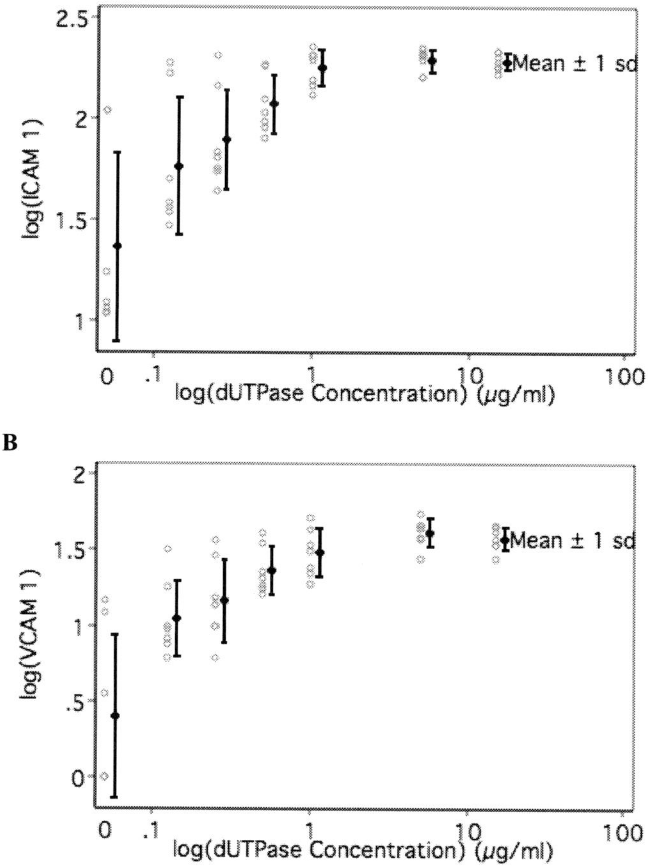

Figure 6A,B. Soluble factors elaborated by EBV-encoded dUTPase-treated macrophages induce inflammation-associated adhesion molecule expression on human endothelial cells. Human umbilical vein endothelial cells were incubated for 8 (VCAM-1) or 24 (ICAM-1) hours with supernatants recovered from EBV dUTPase-treated macrophages, harvested and stained with FITC-conjugated monoclonal antibodies specific for ICAM-1 (A) or VCAM-1 (B), and analyzed by fluorescence flow cytometry. Open symbols represent log mean fluorescence intensity values for each individual donor. Closed symbols represent mean log mean fluorescence intensity values among donors at each dUTPase concentration ± 1 standard deviation.

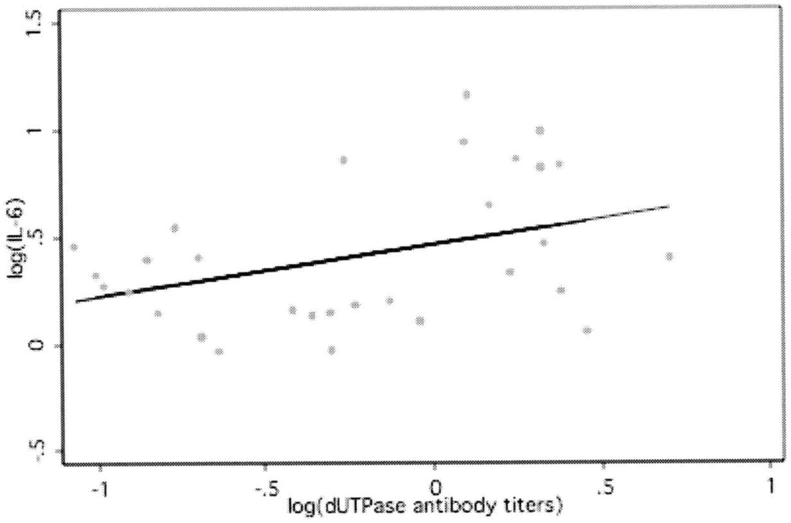

Figure 7. Scatterplot and linear regression line of ln(IL-6) on log(dUTPase). Figure 6A presents a scatter plot of log(IL-6) vs. log(dUTPase) along with the linear regression model (log(IL-6) = 0.464 + 0.243 x log(dUTPase)) based on n = 30 subjects with no missing data and eliminating 2 subjects with log(IL-6) > 3. This association was statistically significant (p = 0.039).

As discussed earlier, antibodies to several EBV-encoded enzymes have been observed in patients with different EBV-associated diseases. To determine whether there was a relationship between anti-EBV-encoded dUTPase antibodies and serum IL-6 levels, a cytokine identified as a co-factor for depression [Kiecolt-Glaser et al., 2003], as well as heart disease [Chen et al., 2008; Fisman et al., 2008] whose expression is up-regulated by the EBV-encoded dUTPase, sera from subjects, who were part of an earlier longitudinal study on caregivers in which plasma IL-6 levels were measured [Kiecolt-Glaser et al., 2003] were examined for antibodies to the EBV-encoded dUTPase. The average age of the 34 men and 60 women in this sample population was 72.18 (SD = 8.92). The subjects' plasma IL-6 values were below 15 pg/ml, with the exception of two participants whose values were 240.58 pg/ml and 831.49 pg/ml, or at least 90 SDs above the group mean; they were excluded from the sample. Both of these subjects were in the group with measurable EBV-encoded dUTPase neutralizing antibody titers; they were removed from the database so as to better conform to the assumptions of equal variance and normality of the IL-6 data within subgroups for the t-test, and because their extreme values could have reflected unreported health problems, e.g., acute illness that was either unknown to the subject or not reported. Subjects with measurable anti-EBV-encoded dUTPase neutralizing antibody levels had generally higher mean IL-6 levels (mean = .41, SEM = 0.06) than did the group whose anti-EBV-encoded dUTPase antibody levels were unmeasurable (mean = 0.34, SEM = 0.03), but the difference between the groups did not achieve statistical significance (p = 0.26). While not statistically significant, the direction of the difference is suggestive. Figure 7 presents a scatterplot of log (plasma IL-6 levels) on log(EBV dUTPase antibody titers) for only those patients with measurable levels of anti-EBV-encoded dUTPase antibody with the two outliers removed. The scatterplot demonstrates that, for those subjects with measurable EBV-encoded dUTPase antibody titers, IL-6 levels tend to increase as the antibody titers increase. This association is significant (p = 0.04), with log (dUTPase) antibody titers explaining r^2 = 14% of the variability in log(IL-6). After controlling for age, the association is still significant (p = 0.047, R^2=14.5%).

THE EBV-ENCODED DUTPASE AND CANCER

EBV is implicated in a variety of human malignancies including BL, NPC, Hodgkin's disease, CLL, non-Hodgkin's lymphoma and gastric cancer [Ablashi et al., 1990; Petrella et al., 1997; Ansell et al., 1999; Brousset et al., 2002; Iwakiri et al., 2003; Thompson and Kurzrock, 2004]. Numerous studies have examined the roles of EBV encoded proteins, such as LMP-1, LMP2A and EBNA 3C, which are expressed during latency, in transformation processes as well as the immunological response of the host to these proteins. However, there have been very few studies directed at determining the role(s) of proteins expressed during the lytic replication in immune modulation or in transformation.

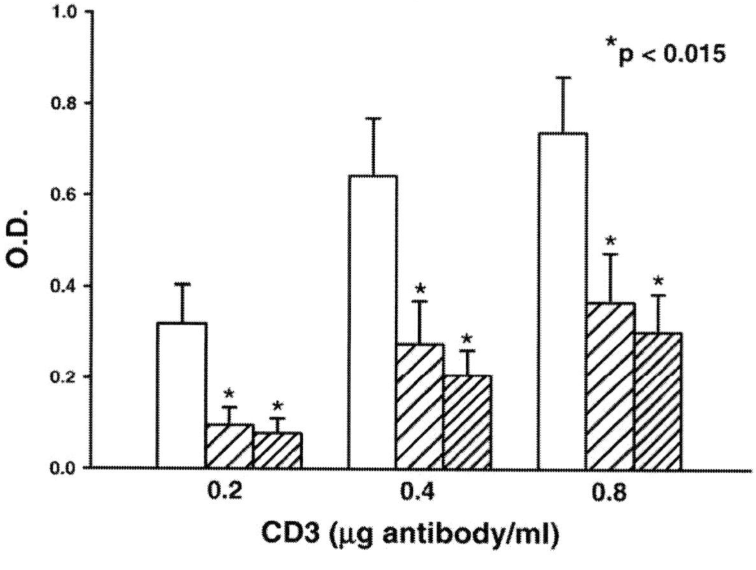

Figure 8A. The effect of EBV dUTPase on the replication of PBMCs stimulated with an anti-CD3 mAb. PBMCs were incubated for 72 hours with various concentrations of an anti-CD3 mAb to induce T-cell blastogenesis. The data are presented as the change in optical density from non-stimulated and CD3-stimulated cells. The PBMCs were treated with EBV dUTPase (cross-hatched bars) at the same time the mAb was added. Results are the mean ± SEM; n=8. *Untreated PBMCs (▭), 7.5 µg/ml EBV dUTPase (▨), and 15 µg/ml EBV dUTPase (▨).*

Cytotoxic T-lymphocytes ($CD8^+$) are responsible for limiting the proliferation and eliminating cells latently infected with EBV [Khanna and Burrows, 2000; Hislop et al., 2007]. However, our knowledge concerning whether T-cells modulate lytic infections [Steven et al., 1997; Khanna and Burrows, 2000; Hislop et al., 2007] or whether viral encoded proteins that are expressed during lytic or abortive infection may modulate the T-cell responses to cells productively or latently infected with EBV is limited. Our initial studies demonstrated that the EBV-encoded dUTPase induced the expression of IL-10, which has been shown to modulate T-cell function and to be importance for establishing persistent viral infections [Barber et al., 2006, Brooks et al., 2006]. To determine whether the EBV-encoded dUTPase affected T-cell function PBMCs from eight individuals were treated with anti-CD3 monoclonal antibody and with or without the EBV-encoded dUTPase and with anti-CD3 monoclonal antibody and 72 hr later proliferation was determined (Figure 8). Treatment of the PBMCs with EBV-encoded dUTPase resulted in a significant inhibition of proliferation of the T-cells when compared to cells only treated only with the anti-CD3 mAb. Using PBMCs from three of these subjects, we found that inhibition of the proliferation response of T-cells by EBV-encoded dUTPase was dose-dependent with saturation occurring at approximately 1 μg/ml.

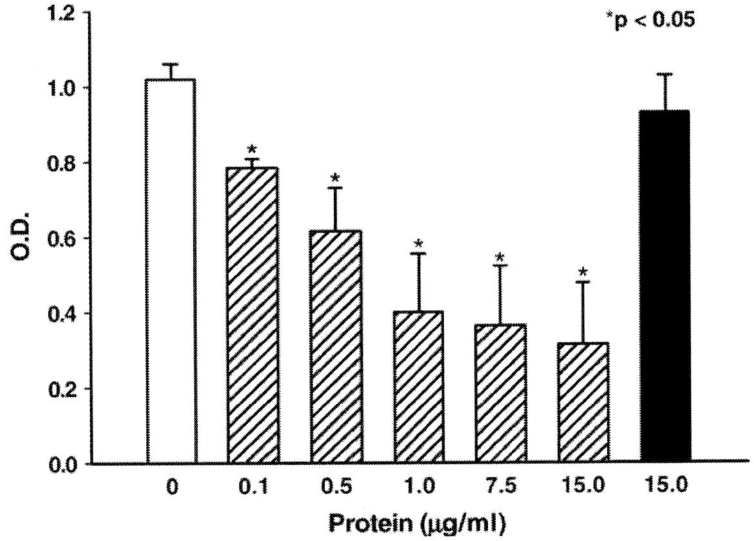

Figure 8B. Dose-dependent inhibition of the replication of PBMCs stimulated with an anti-CD3 mAb by EBV dUTPase. PBMCs were incubated with 0.8 μg/ml of an anti-CD3 mAb for 72 hours as described in the Methods. The cells were treated with various concentrations of EBV dUTPase. The results are shown as the mean ± SEM; n=3. *Untreated PBMCs (▭), EBV dUTPase (▨), and human γ-globulin (▩).*

To evaluate whether the EBV-encoded dUTPase was capable of suppressing an EBV-specific T-cell response, we performed an experiment to determine whether the EBV-encoded dUTPase had an effect on the transformation of B-cells. Briefly, PBMCs were infected with the B95-8 transforming strain of EBV in the presence and absence of the dUTPase. While transformation of resident B-cells normally occurs the proliferation and expansion of this B-cell population is limited by the presence of EBV specific CTLs. However, as can be seen in Figures 9 and 10, B-cell transformation was enhanced 6-and 10-fold ($p < 0.001$) when treated with 0.1 and 1.0 μg/ml dUTPase respectively when compared to cells just treated with virus. These data suggest that the dUTPase is preventing killing of the B-cells by EBV specific CTLs. These results suggest that the expression of the EBV-encoded dUTPase might represent a possible mechanism by which the virus can establish and maintain persistence in the host and perhaps, prevent effective immune surveillance mechanisms from eliminating EBV-positive transformed cells.

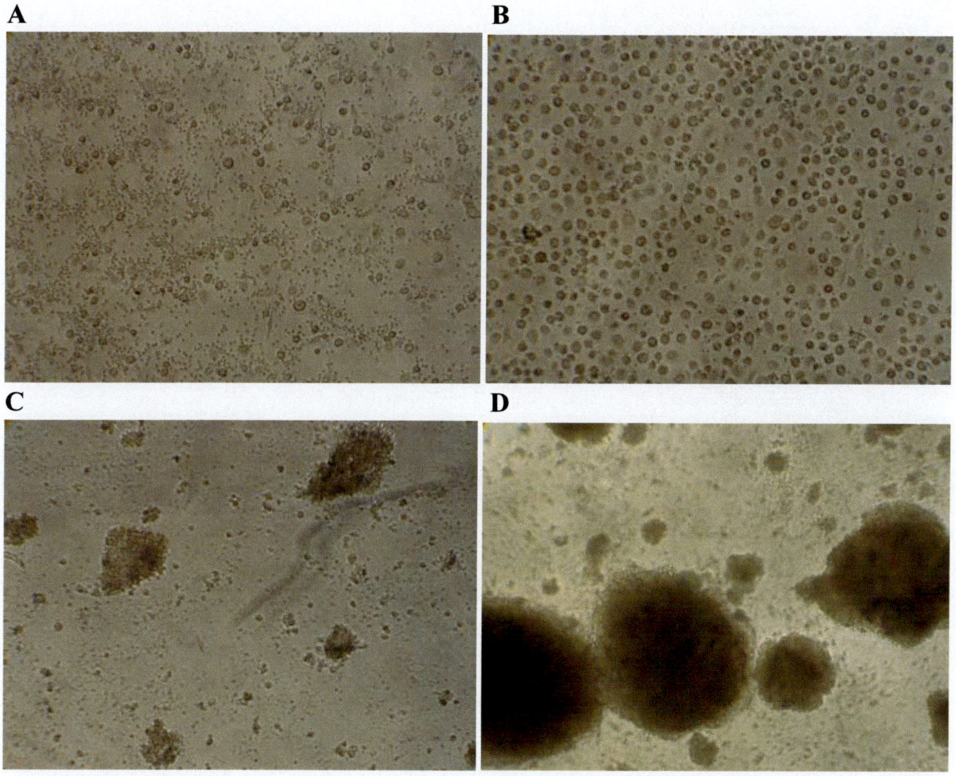

Figure 9. The EBV-encoded dUTPase enhances transformation by the B95 strain of EBV. PBMCs (7 x 10^5 cells/ml) were incubated with 10^6 transforming units of the B95-8 transforming strain of EBV. If present, the EBV-encoded dUTPase was added at the same time as virus addition. The EBV-encoded dUTPase was replenished every 3 days for two weeks. After 28 days, the cells were examined. A. Untreated controls B. B95-treated. C B95 and dUTPase (0.1 μg/ml) D. B95 and dUTPase treated (1 μg/ml).

Conclusion

The innate and adaptive immune responses are critical processes involved with the detection and clearance of invading pathogens. Pattern recognition receptors (PRR), which include TLRs, have been demonstrated to play an important role in modulating innate recognition of viruses by not only serving as pathogen sensors but also by activating signaling pathways that result in the increased production of various cytokines and chemokines, which activate cells of the innate immune system and stimulate the adaptive immune system. To better understand the immune responses triggered by viruses, it is critical to identify not only what PRRs are activated during viral infection but also the specific viral components that act as ligands/triggers for the different PRRs leading to the stimulation of the host immune response.

The data from these studies show that that treatment of human monocytes derived macrophages with the EBV-encoded dUTPase led to the increased expression and secretion of TNF-α, IL-1β, IL-6, IL-8, as well as, IL-10 and that the induction of these cytokines was dependent upon NF-κB, which was activated through a TLR2 dependent pathway [Glaser et al., 2006; Waldman et al, 2007; Ariza et al., 2008]. These data support the hypothesis that proteins that compose the EBV EA complex can modulate immune function. Recent studies, which also support this hypothesis, have shown that in addition to it endonuclease activity the EBV-encoded DNase (BGLF5), another protein in the EA complex, also functions to shut down host protein synthesis, which results in blocking the synthesis of HLA class 1 and II molecules [Rowe et al., 2007; Zuo et al., 2008].

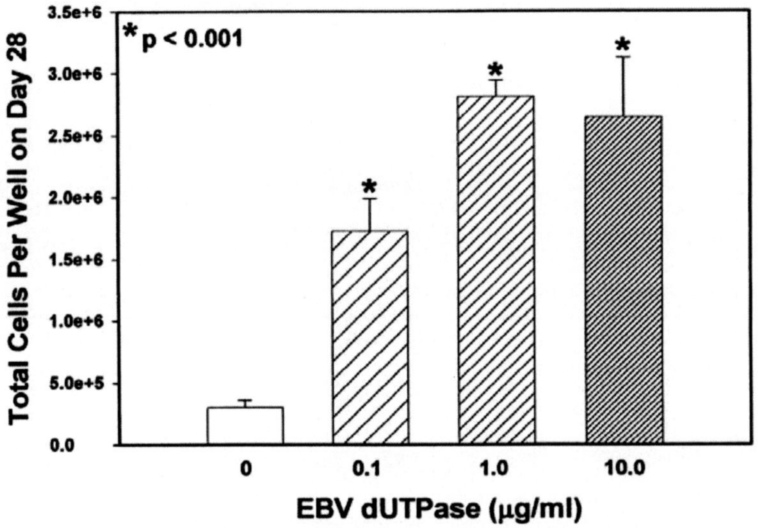

Figure 10. The EBV-encoded dUTPase enhances transformation by the B95 strain of EBV. PBMCs (7 x 10^5 cells/ml) were incubated with 10^6 transforming units of the B95-8 transforming strain of EBV. If present, the EBV-encoded dUTPase was added at the same time as virus addition. The EBV-encoded dUTPase was replenished every 3 days for two weeks. After 28 days, the cells were examined. The average number of cells ± SD over four wells was determined for each treatment.

Two important questions that arise from these studies are: does the EBV-encoded dUTPase have a role(s) in virus replication in addition to its role in viral DNA synthesis and does the EBV encoded dUTPase contribute to the pathophysiology observed in infections caused by EBV? We believe that the data suggests that the dUTPase may play a role in EBV replication by inducing the increased expression of pro-inflammatory cytokines, which attract cells that could become infected by the virus and possibly result in its dissemination to other sites within the host [Salek-Ardakani, et al., 2004; Tugizov, et al., 2007].

Our data also demonstrates that the EBV-encoded dUTPase may contribute directly to the pathophysiology observed in infections caused by EBV. As already discussed, inflammation plays a central role in driving the evolution of atherosclerosis and coronary artery disease. Not only does the dUTPase induce pro-inflammatory cytokines but it also cause the upregulation in the surface expression of inflammation –associated endothelial adhesion molecules VCAM-1 and ICAM-1, providing further evidence that the EBV-encoded dUTPase induces an inflammatory cascade. Furthermore, there was a direct correlation between IL-6 levels and anti-dUTPase antibody levels. While sera from normal EBV-seronegative or seropositive individuals do not contain anti-dUTPase antibodies [Sommer et al., 1996; our unpublished data], we have detected anti-dUTPase antibodies sera from caregiver subjects (34%; n= 94) [Waldman et al., 2007]; in patients with nasopharyngeal carcinoma (44%; n = 16) and in patients with chronic fatigue syndrome (CFS) (15%; n = 67). Further studies are necessary to determine the relationship, if any, between the presence of neutralizing antibody to the EBV-encoded dUTPase and pathogenesis.

Perhaps more importantly, the EBV-encoded dUTPase could represent a mechanism to dampen the innate and adaptive immune responses to the virus. We have demonstrated that T –cell function is suppressed in mice treated with the EBV-encoded dUTPase and that these mice exhibit sickness behavior similar to that observed in patients with CFS [Padgett et al., 2004]. Furthermore, we have shown that the EBV-encoded dUTPase stimulates cytokine IL-10 production in human monocytes derived macrophages [Glaser et al., 2006]. IL-10 affects $CD8^{+}$ T cell function and its increased expression is a primary evasion mechanism used by viruses to establish persistent infections [Brooks et al., 2006; Ejrnaes et al., 2006] and in tumor progression [Tanner et al., 1997; Khatri and Caliguiri, 1998; Fayad et al., 2001; Lai et al., 2002 Budiani et al., 2002;].

It is clear that additional studies are necessary to define the roles of the EBV-encoded dUTPase virus replication and in the pathophysiology of disease. However, studies should be expanded to include not only other proteins of the EA complex of EBV, but also homologous proteins in other human herpesviruses. Likewise, studies need to be performed to determine the relationship between antibodies to the protein and disease. Such studies may provide diagnostic methods as well as developing new therapeutic approaches that could be used to treat diseases caused by these viruses.

References

Ablashi DV, Huang AT, Pagano JS, Pearson GR and Yang CS. 1990. *Epstein-Barr Virus and Human Disease*. Humana Press, Clifton NJ.

Akira, S., and K. Takeda. 2004. Toll-like receptor signaling. *Nature Immunol. 4*:499-511.

Ansell, S.M., Li, C.-Y., Lloyd, R.V., and Phyliky, R.L., 1999. Epstein–Barr virus infection in Richter's transformation. *Am. J. Hematol. 60*: 99– 104.

Anatomy H, Doucette R, Bearing MT, Pisa P, Kissing R, and Masco MG. 1994. The Epstein-Barr virus latent membrane protein-1 (LMP1) induces interleukin-10 production in Burkett lymphoma lines. *Int J Cancer 57*: 240-244.

Aravalli, R.N., S. Hu, T.N. Rowen, J.M. Palquist, and J.R. Lokensgard. 2005. TLR2-mediated proinflammatory cytokine and chemokine production by microglial cells in response to herpes simplex virus. *J. Immunol. 175*: 4189-4193.

Ariza ME, Glaser R, Kaumaya PTP, Jones C and Williams MV. 2008. The Epstein-Barr Virus-Encoded dUTPase Activates NF-κB through the TLR2 and MyD88-dependent Signaling Pathway. Submitted.

Asea, A., M. Rehli, E. Kabingu, J.A. Boch, O. Bare, P.E. Auron, M.A. Stevenson, and S.K. Calderwood. 2002. Novel signal transduction pathway utilized by extracellular HSP70: role of toll-like receptor (TLR) 2 and TLR 4. *J. Biol. Chem. 277*:15028-15034.

Baldo AM and McClure M. 1999. Evolution and horizontal transfer of dUTPase encoding genes in viruses and their hosts. *J. Virol 73*:7710-7721.

Barber DL, Wherry EJ, Masopust D, Zhu B, Allison JP, Sharpe AH, Freeman GJ and Ahmed R. 2006. Restoring function in exhausted $CD8^+$ T cells during chronic viral infection. *Nature 439*:682-687.

Beaulieu AD, Paquin R., and Gosselin J. 1995. Epstein-Barr virus modulates de novo protein synthesis in human neutrophils. *Blood 86*:2789-2798.

Bernier-Villamor V, Camacho A, Hidalgo-Zarco, F, Perez J, Ruiz-Perez LM and Gonzalez-Pacanowska, D. 2002 Characterization of deoxyuridine triphosphate nucleotidohydrolase from *Trypanosoma cruzi. FEBS Lett. 526*:147-150.

Beutler B. 2004. Interferences, questions and possibilities in Toll-like receptor signaling. *Nature. 430*:257-263.

Binder, C.J., Chang, M.K., Shawn, P.X., Miller, Y.I., Hartvigsen, K., Dewan, A., Witztum, J.L., 2002. Innate and acquired immunity in atherogenesis. *Nat Med 8*, 1218-1226.

Blasig C, Zietz C, Haar B, Neipel F, Esser S, Brockmeyer NH, Tschachler E, Colombini S, Ensoli B and Sturzl M. 1997. Monocytes in Kaposi's sarcoma lesions are productively infected by human herpes virus 8. *J. Virol. 71*:7963-7968.

Bonnet M, Guinebretiere JM, Kremmer E, Grunewald V, Benhamou E, Contesso G and Jaob I. 1999. Detection of Epstein-Barr virus in invasive breast cancers. *J Natl Cancer Inst 91*:1376-1381.

Brooks DG, Trifilo MJ, Edelmann KH, Teyton L, McGavern DB and Oldstone MBA. 2006. Interleukin-10 determines viral clearance or persistence *in vivo*. *Nature Med 12*:1301-1309.

Brousset P. 2002. Is the Epstein-Barr virus infection relevant in lympomagenesis? *Human Pathol. 33*:143-145.

Budiani DR, Hutahaean S, Haryana SM, Soesatyo M and Sosroseno, W. 2002. Interleukin-10 levels in Epstein-Barr virus-associated nasopharyngeal carcinoma. *J Microbiol Immunol Infect 35*:265-271.

Camacho A, Arrebola R, Pena-Diaz J, Ruiz-Perez LM, and Gonzalez-Pacanowska D. 1997. Description of a novel eukaryotic deoxyuridine-5'-triphosphate nucleotidohydrolase in *Leishmania major*. *Biochem. J 325*: 441-447.

Camacho A, Hidalgo-Zarco F, Bernier-Villamor V and Ruiz-Rerez LM. 2000. Properties of *Leshmania major* dUTP nucleotidohydrolase, a distinct nucleotide-hydrolyzing enzyme in kinetoplasts. *Biochem. J. 346*:163-168.

Chen D, Assad-Kottner C Orrego C and Torre-Amigone G. 2008. Cytokines and acute heart failure. *Crit Care Med 36 (Suppl 1)*: S9-16.

Cheng, Y-C., Chen, Y.Y., Glaser, R., and Henle, W., 1980a. Frequency and levels of antibodies to Epstein–Barr virus specific DNase are elevated in patients with nasopharyngeal carcinoma. *Proc. Natl. Acad. Sci. USA 77 (10)*: 6162–6165.

Cheng, Y.-C., Chen, J.Y., Hoffman, P.J., Glaser, R. 1980b. Studies on the activity of DNase associated with the replication of the Epstein–Barr virus. *Virology 100*: 334–338.

Clough, W. 1979. Deoxyribonuclease activity found in Epstein-Barr virus producing lymphoblastoid cells. *Biochemistry 18*: 4517-4521.

D'Addario M, Ahmad A, Xu JW, and Menezes J. 1999. Epstein-Barr virus envelope glycoprotein gp350 induces NF-κB activation and IL-1β synthesis in human monocytes-macrophages involving PKC and PI3-K. *FASEB J. 13*:2203-2213

D'Addario M, Ahmad A, Morgan A and Menezes J. 2000. Binding of the Epstein-Barr virus major envelope glycoprotein gp350 results in the upregulation of the TNF-αgene expression in monocytic cells via NF-κB involving PKC, P13-K and tyrosine kinases. *J Mol Biol 298*:765-778.

D'Addario M, Libermann TA, Xu J, Ahmad A, and Menezes J. 2001. Epstein-Barr virus and its glycoprotein-350 upregulate IL-6 in human B-lymphocytes via CD21, involving activation of NF-κB and different signaling pathways. *J Mol Biol 308*: 501-514.

Danesh, J., and Appleby, P. 1998. Persistent infection and vascular disease: a systematic review. *Expert Opin Investig Drugs 7*: 691-713.

Danesh, J., Collins, R., and Peto, R. 1997. Chronic infections and coronary heart disease: Is there a link? *Lancet 350*: 430-436.

Davidson AJ and Stow ND. 2005. New genes from old: redeployment of dUTPase by herpesviruses. *J Virol 79*:12880-12892.

de Boer, O.J., Teeling, P., Idu, M.M., Becker, A.E., van der Wal, A.C., 2006. Epstein Barr virus specific T-cells generated from unstable human atherosclerotic lesions: Implications for plaque inflammation. *Atherosclerosis 184*: 322-329.

de Sanjose S, Bosch R, Schouten T, Verkuijlen S, Nieters N, Brennan P, Benavente Y, Boffetta P, Meijer CJLM and Middeldorp JM. 2007. Epstein-Barr virus infection and risk of lymphoma: immunoblot analysis of antibody responses against EBV-related proteins in a large series of lymphoma subjects and matched controls. *Int J Cancer 121*:1806-1812.

Deng JH, Zhang YJ, Wang X. and Gao SJ. 2004. Lytic replication-defective Kaposi's sarcoma-associated herpesvirus; potential role in infection and malignant transformation. *J. Virol 78*:11108-11120.

Dukers DF, Meij P, Vervoort MB, Vos W, Scheper RJ, Meijer CJ, Bloemena E and Middeldorp JM. 2000. Direct immunosuppressive effects of EBV-encoded latent membrane protein 1. *J. Immunol. 165* (2): 663–670.

Ejrnaes M, Filippi CM, Martinic MM, Ling EM, Togher LM, Crotty S and von Herrath MG. 2006. Resolution of a chronic viral infection after interleukin-10 receptor blockade. *J Infect Dis. 203*:2461-2472.

El-Hajj HH, Zhang H and Weiss B. 1988. Lethality of a *dut* (deoxyuridine triphosphatase) mutation in *Escherichia coli*. *J Bacterial 170*:1069-1075.

Eliopoulos AG, Gallagher NJ, Blake SM, Dawson CW, and Young LS. 1999. Activation of the p38 mitogen-activated protein kinase pathway by Epstein-Barr virus-encoded latent membrane protein 1 coregulates interleukin-6 and interleukin-8 production. *J Biol Chem.274*: 16085-16096.

Esterling BA, Antoni MH, Kumar M, and Schneiderman N. 1990. Emotional repression, stress disclosure responses, and Epstein-Barr viral capsid antigen titers. Psychosom. *Med. 52*:397-410.

Esterling BA, Antoni MH, Schneiderman N, Carver CS, LaPerriere A, Ironson G. Klimas NG and Fletcher MA. 1992. Psychosocial modulation of antibody to Epstein-Barr viral capsid antigen and human herpesvirus type-6 in HIV-infected and at-risk gay men. *Psychosom. Med. 54*:354-371.

Fayad L, Keating MJ, Reuben JM, O'Brien, S, Lee B-N, Lerner S and Kurzrock R. 2001. Interleukin-6 and interleukin-10 levels in chronic lymphocytic leukemia: correlation with phenotypic characteristics and outcome. *Blood 97*:256-263.

Fiorentino DF, Bond MW and Mosmann TR. 1989. Two types of mouse T helper cell IV. Th2 clones secrete a factor that inhibits cytokine production by Th1 clones. *J. Exp. Med. 170*:2081-2095.

Finberg RW, Wang JP and Kurt-Jones EA. 2007. Toll like receptors and viruses. *Rev Med Virol 17*: 35-43.

Fisman EZ, Adler T and Tenenbaum A. 2008. Biomarkers in cardiovascular diabetology: interleukins and matrixins. *Adv Cardiol. 45*: 44-64.

Fleischmann J, Kremmer E, Greenspan JS, Grasser FA and Niedobitek G. 2002. Expression of viral and human dUTPase in Epstein-Barr virus-associated diseases. *J Med Virol 68*:568-573.

Gadsden MH, McIntosh EM, Game JC, Wilson PJ and Haynes RH. 1993. dUTP pyrophosphatase is an essential enzyme in *Saccharomyces cerevisiae*. *EMBO J 12*:4425-4431.

Gaudreault E., S. Fiola, M. Olivier, and J. Gosselin. 2007. Epstein-Barr virus induces MCP-1 secretion by human monocytes via TLR2. *J Virol. 81*:8016-8024.

Glaser R, Ogino T, Zimmerman Jr J and Rapp F. 1973. Thymidine kinase activity in Burkitt lymphoblastoid somatic cell hybrids after induction of the EB virus. *Proceedings of the Society for Experimental Biology and Medicine 142*: 1059-1062.

Glaser R, de-Thé G, Lenoir G and Ho JHC. 1976. Superinfection of epithelial nasopharyngeal carcinoma cells with Epstein-Barr virus. *Proc Natl Acad Sci USA 73*: 960-963.

Glaser R, Lang CM, Lee KJ, Schuller DE, Jacobs D and McQuattie C. 1980. Attempt to infect nonmalignant nasopharyngeal epithelial cells from humans and squirrel monkeys with Epstein-Barr virus. *J. Natl Cancer Inst. 64*: 1085-1090.

Glaser R, Rice J, Sheridan J, Fertel R, Stout J, Speicher C, Pinsky D, Kotur M, Post A., and Beck M. 1987. Stress-related immune suppression: health implications. *Brain Behav Immun. 1*:7-20.

Glaser R, Pearson, GR, Jones JF, Hillhouse J, Kennedy S, Mao HY, and Kiecolt-Glaser JK. 1991. Stress-related activation of Epstein-Barr virus. *Brain Behav Immun. 5*:219-232.

Glaser R, Pearson GR, Bonneau RH, Esterling BA, Atkinson C, and Kiecolt-Glaser, JK. 1993. Stress and the memory T-cell response to the Epstein-Barr virus in healthy medical students. *Health Psychology. 12*:435-442.

Glaser R, and Kiecolt-Glaser JK. 1994. Stress-associated immune modulation and its implications for reactivation of latent herpesviruses. In: Glaser R, Jones, J., ed. *Human Herpesvirus infections*. New York: Marcel Dekker; 1994:245-270.

Glaser R and Kiecolt-Glaser J.K. 2005. Stress-induced immune dysfunction: Implications for health. *Nature Reviews Immunol. 5*:10-18.

Glaser R, Litsky, ML, Padgett DA, Baiocchi RA, Yang EV, Chen M, Yeh PE, Klimas NG, Marshall GD, Whiteside T, Herberman R, and Williams MV. 2005. Stress-associated changes in the steady state expression of latent Epstein-Barr virus: Implications for Chronic Fatigue Syndrome and Cancer. *Brain Behav Immun. 19*:91-103

Glaser R, Litsky ML, Padgett DA, Baiocchi RA, Yang EV, Chen M, Yeh PE, Green-Church KB, Caligiuri MA and Williams MV. 2006. The EBV-encoded dUTPase induces immune dysregulation: Implications for the pathophysiology of EBV-associated disease. *Virology 346*:205-218.

Gosselin J., J. Menezes, M. D'Addario, J. Hiscott, L. Flamand, G. Lamoureux, and D.Oth. 1991. Epstein-Barr virus inhibits tumor necrosis factor-α transcription in monocytic cells. *Eur J Clin Immunol. 21*:203-205.

Gosselin J., L. Flamand, M. D'Addario, J. Hiscott, and J. Menezes. 1992. Infection of peripheral blood mononuclear cells by herpes simplex and Epstein-Barr viruses. Differential induction of interleukin 6 and tumor necrosis factor-α. *J. Clin. Invest. 89*:1849-1856.

Gosselin J., L. Flamand, M. D'Addario, J. Hiscott, I. Stefanescu, D.V. Ablashi, R.C. Gallo, and J. Menezes. 1992. Modulatory effects of Epstein-Barr, herpes simplex and human herpes-6 viral infections and co-infections on cytokine synthesis: a comparative study. *J. Immunol. 149*:181-187.

Gosselin J and Borgeat P. 1997. Epstein-Barr virus modulates 5-lipogenase product synthesis in human peripheral blood mononuclear cells. *Blood 89*:2122-2130.

Gosselin J, Savard M, Tardif M, Flamand L and Borgeat P. 2001. Epstein-Barr virus primes human polymorphonuclear leukocytes for the biosynthesis of leukotriene B4. *Clin Exp Immunol 126*:494-502.

Hahn AM, Huye LE, Ning S, Webster-Cyriaque J and Pagano JS. 2005. Interferon regulatory factor 7 is negatively regulated by the Epstein-Barr virus immediate-early gene BZLF-1. *J. Virol 79*: 10040-10052

Henle G, Henle W and Klein G. 1971. Demonstration of two distinct components in the early antigen complex of Epstein-Barr virus infected cells. *Int J Cancer 8*: 272-282.

Henle W, Ho JHC, Henle G, and Kwan HC. 1973. Antibodies to Epstein-Barr virus-related antigens in nasopharyngeal carcinoma. Comparison of active cases and long-term survivors. *Journal of the National Cancer Institute 51*: 361-369.

Henry BE, Glaser R, Hewetson J, and O'Collaghan, D. 1978. Expression of altered ribonucleotide reductase activity associated with the replication of the Epstein–Barr virus. *Virology 89*: 262–271.

Hill F, Loakes, D and Brown DM. 1998. Polymerase recognition of synthetic oligodeoxyribonucleotides incorporating degenerate pyrimidine and purine bases. *Proc. Natl. Acad. Sci. USA. 95*: 4258-4263.

Hislop AD, Taylor GS, Sauce D and Rickinson AB. 2007 Cellular responses to viral infection in humans: lessons from Epstein-Barr virus. *Annu. Rev. Immunol. 25*: 587-617.

Ho M, Jaffe R, Miller G, Breinig MK, Dummer JS, Makowka L, Atchison RW, Karrer F, Nalesnik MA and Starzl TE. 1988. The frequency of Epstein-Barr virus infection and associated lymphoproliferative syndrome after transplantation and its manifestations in children. *Transplantation 45*:719-728.

Horvath R, Cerny J, Benedik Jr J, Hokl J, Jelinkova I, and Benedik J. 2000. The possible role of human cytomegalovirus (HCMV) in the origin of atherosclerosis. *J Clin Virol 16*, 17-24.

Huang J, Chen H, Hutt-Fletcher L, Ambinder RF and Hayward SD. 2003. Lytic viral replication as a contributor to the detection of Epstein-Barr virus in breast cancer. *J Virol 77*:13267-13274.

Ibrahim, A.I., Obeid, M.T., Jouma, M.J., Moasis, G.A., Al-Richane, W.L., Kindermann, I., Boehm, M., Roemer, K., Mueller-Lantzsch, N., Gartner, B.C., 2005. Detection of herpes simplex virus, cytomegalovirus and Epstein-Barr virus DNA in atherosclerotic plaques and in unaffected bypass grafts. *J Clin Virol 32*, 29-32.

Iwakiri D, Eizuru Y, Tokunaga M and Takada, K. 2003. Autocrine growth of Epstein-Barr virus positive gastric carcinoma cells mediated by Epstein-Barr virus-encoded small RNA. *Cancer Res. 63*:7062-7067.

Jones, J.F., Williams, M.V., Schooley, R.T. and Glaser, R. 1988. Antibodies to Epstein-Barr virus specific DNase and DNA polymerase in the chronic fatigue syndrome. *Arch. Intern. Med. 148*: 1957-1960.

Kanda K, Kempkes B, Bornkamm GW, von Gabain A, and Decker T. 1999. The Epstein-Barr virus nuclear antigen 2 (EBNA2), a protein required for B lymphocyte immortalization, induces the synthesis of type I interferon in Burkitt's lymphoma cell lines. *Biol Chem. 380*:213-221.

Kawai T., and S. Akira S. 2006. Innate immune recognition of viral infection. *Nature Immunol. 7*:131-137.

Keiff E and Rickinson AB. Epstein Barr virus and its replication. In B.N. Fields, D.M. Knipe, and P.M. Howley (eds.) *Fields Virology*, Ed 4 Vol 2 pp.2511-2575. Philadelphia, PA: Lippincott-Raven. 2001.

Khanna R and Burrows SR. 2000. Role of cytotoxic T lymphocytes in Epstein-Barr virus-associated diseases. *Annu Rev Microbiol 54*:19-48.

Khatri VP and Caliguiri MA. 1998. A review of the association between interleukin-10 and human B-cell malignancies. *Cancer Immunol Immunother 46*:239-244.

Kiecolt-Glaser J, Preacher K, MacCallum R, Atkinson C, Malarkey W and Glaser R.,2003. Chronic stress and age-related increases in the proinflammatory cytokine IL-6. Proc *Nat Acad Sci USA 100*: 9090-9095.

Klein SC, Kube D, Abts H, Diehl V and Tesch H. 1996. Promotion of IL8, IL10, TNF-α and TNF- βproduction by EBV infection. *Leuk Res 20*:633-636.

Kurt-Jones, EA, Belko J, Yu C, Newburger PE, Wang J, Chan M, Knipe DM, and Finberg RW. 2005. The role of Toll-like receptors in herpes simplex infection of neonates. *J. Infect. Dis. 191*:746-748.3.

Kurzrock R. 2001. Cytokine deregulation in cancer. *Biomed Pharmacother 55*:543-547.

Lai R, O'Brien S, Maushouri T, Rogers A, Kantarjian H, Keating M and Albitar, M. 2002. Prognostic value of plasma interleukin-6 levels in patients with chronic lymphocytic leukemia. *Cancer 95*:1071-1075.

Laichalk LL and Thorley-Lawson DA. 2005. Terminal differentiation into plasma cells initiates the replicative cycle of Epstein-Barr virus *in vivo*. *J Virol* 1296-1307.

Larochelle B, Flamand L, Gourde P, Beauchamp D and Gosselin J. 1998. Epstein-Barr virus infects and induces apoptosis in human neurophils. *Blood 92*:291-299.

Levine PH, Ablashi DV, Pearson GR and Kottaridis SD. 1985. *Epstein-Barr Virus and Associated Diseases*. Martinus Nijhoff Publishing, Boston MA.

Liu MY, Cou WM, Nutter L, Hsu MM and Chen, JY.1989. Antibody against Epstein-Barr virus DNA polymerase activity in sera of patients with nasopharyngeal carcinoma. *Journal of Medical Virology 28*: 101-105.

Lu CC, Huang HT, Wang JT, Slupphaug G, Li TK, Wu MC, Chen YC, Lee CP and Chen MR. 2007. Characterization of the uracil-DNA glycosylase activity of Epstein-Barr virus BKRF3 and its role in lytic virus DNA replication. *J. Virol. 81*:1195-1208.

Luciani M G, Stoppacciaro A, Peri G, Mantovani A and Ruco LP. 1998. The monocyte chemotactic protein α (MCP-1) and interleukin 8 (IL-8) in Hodgkin's disease and in solid tumours. *Mol Pathol 51*:273-278.

Lund J., A. Sato, S. Akira, R. Medzhitov, and A. Iwasaki. 2003. Toll-like receptor 9-mediated recognition of herpes simplex virus-2 by plasmacytoid dendritic cells. *J Exp Med. 198*:513-520.

Mansur, D.S., E.G. Kroon, M.L. Nogueria, R.M.E. Arantes , S.C.O. Rodrigues, S. Akira, R.T. Gazzinelli, and M.A. Campos. 2005. Lethal encephalitis in myeloid differentiation factor 88-deficent mice infected with herpes simplex virus 1. *Am. J. Patho. 166*:1419-1426.

Martin H.J., J.M. Lee, D. Walls, and S.D. Hayward. 2007. Manipulation of the Toll-like receptor 7 signaling pathway by Epstein-Barr virus. *J. Virol. 81*:9748-9758.

Masy E, Adriaenssens E, Montpellier C, Crepiex P, Mougel A, Quatannens B, Goormachtigh G, Faumount N, Meggetto F, Auriault C, Groux H and Coll J. 2002. Human monocytic cell lines transformed *in vitro* by Epstein-Barr virus display a type II latency and LMP-1-dependent proliferation. *J Virol 76*:6460-6472.

McColl, S.R., Roberge, C.J., Larochelle, B., Gosselin, J., 1997. EBV induces the production and release of IL-8 and macrophage inflammatory protein-1 alpha in human neutrophils. *Journal of Immunology 159*, 6164-6168.

McGeoch DJ, Cook S, Dolan A, Jamieson FE and Telford EAR. 1995. Molecular phylogeny and evolutionary timescales for the family of mammalian herpesviruses. *J Mol Biol 247*:443-458

Miller R, Glaser R and Rapp F. 1977. Studies of an Epstein-Barr virus-induced DNA polymerase. *Virology 76*: 494-502.

Miyazaki I, Cheung RK and Dosch H-M. 1993. Viral interleukin 10 is critical for the induction of B cell growth transformation by Epstein-Barr virus. *J. Exp. Med. 178*:439-447.

Mori A, Takao S, Pradutkanchana J, Kietthubthew S, Mitarnun W and Ishida T. 2003. High tumor necrosis factor-alpha levels in patients with Epstein-Barr virus-associated peripheral T-cell proliferative disease/lymphoma. *Leuk Res 27*:493-498.

Morrison TE, Mauser A, Klingelhutz A and Kenney SC. 2004. Epstein-Barr virus immediate-early protein BZLF1 inhibits tumor necrosis factor alpha-induced signaling and apoptosis by downregulating tumor necrosis factor receptor 1. *J. Virol. 78*: 544-549.

Moskophidis D., Lechner F, Pircher H and Zinkernagel RM. 1993. Viral persistence in acutely infected immunocompetent mice by exhaustion of antiviral cytotoxic effector T cells. *Nature 362*:758-761.

Nakamura H, Iwakiri D, Ono Y and Fujiwara S. 1998.Epstein-Barr virus-infected human T cell line with a unique pattern of viral gene expression. *Int J Cancer* 76:587-594.

Nicholls JM, Sommer P, Kremmer E, Ong KS, Fung K, Lee JMF, Ng, MH and Grasser FA. 1998. A new lytic antibody, 7D6, detects Epstein-Barr virus dUTPase in nonkeratinizing undifferentiated nasopharyngeal carcinomas. *Lab Invest* 78:1031-1032.

Nyman PO. 2001.dUTPases introduction. *Curr Protein Peptide Sci 2*:277-287.

O'Garra A and Vieira P. 2007. T_H1 cells control themselves by producing interleukin-10. *Nature Rev Immunol* 7:425-428.

Padgett DA, Hotchkiss AK, Pyter LM, Nelson RJ, Yang E, Yeh PE, Litsky M, Williams M, and Glaser R. 2004. Epstein-Barr virus-encoded dUTPase modulates immune function and induces sickness behavior in mice. *J Med Virol 74*:442-448.

Paramita DK, Fachiroh J, Artama WT, van Benthem E, Haryana SM and Middleldrof JM. 2007. Native early antigen of Epstein-Barr virus, a promising antigen for diagnosis of nasopharyngeal carcinoma. *J Med Virol 79*:1710-1721.

Parkhill J, Wren BW, Mungall K, Ketley JM, Churcher C, Basham D, Chillingworth T, Davies RM, Feltwell T, Holroyd S, Jagels K, Karlyshev AV, Moule S, Pallen MJ, Penn CW, Quail MA, Rajandream MA, Rutherford KM, van Vliet A.H.M, Whitehead S and Barrell BG. 2000. The genome sequence of the food-borne pathogen *Campylobacter jejuni* reveals hypervariable sequences. *Nature 403*: 665-669.

Pegtel DM, Middleldorp J and Thorley-Lawson DA. 2004. Epstein-Barr virus infection in *ex vivo* tonsil epithelial cell cultures of asymptomatic carriers. *J Virol 78*:12613-12624.

Petrella T, Yaziji N, Colin F, Rifle G, Morlevat F, Arnould L, Fargeot P and Depret O. 1997. Implication of the Epstein-Barr virus in the progression of chronic lymphocytic leukaemia/small lymphocytic lymphoma to Hodgkin-like lymphomas. *Anticancer Research 17*:3907-3913.

Pober, J.S., Cotran, R.S., 1990. Cytokines and endothelial cell biology. *Physiological Reviews 70*, 427-451.

Rattazzi M, Faggin E, Pertipaglia B and Pauletto P. 2005. Innate immunity and atherogenesis. *Lupus 14*, 747-751.

Reedman BM and Klein G. 1973. Cellular localization of an Epstein-Barr Virus (EBV)-associated complement-fixing antigen in producer and non-producer lymphoblastoid cell lines. *Int J Cancer 11:* 599-620.

Ridker PM, Cannon CP, Morrow D, Rifai N, Rose LM, McCabe CH, Pfeffer MA and Braunwald E. 2005a. C-reactive protein levels and outcomes after statin therapy. *N Engl J Med 352*: 20-28.

Ridker PM, Rifai N, Cook NR, Bradwin G, and Buring, JE, 2005b. Non-HDL cholesterol, apolipoproteins A-I and B100, standard lipid measures, lipid ratios, and CRP as risk factors for cardiovascular disease in women. *JAMA 294*: 326-333.

Rowe M., Glaunsinger B, van Leeuwen D, Zuo J, Sweetman D, Ganem D, Middeldorp J, Wiertz EJHJ and Ressing ME. 2007. Host shutoff during productive Epstein-Barr virus infection is mediated by BGLF5 and may contribute to immune evasion. *Proc Natl Acad Sci USA 104*: 3366-3371.

Salek-Ardakani S, Lyons LA and Arrand, JR. 2004.Epstein-Barr virus promotes human monocyte survival and maturation through a paracrine induction of INF-α. *J Immunol. 173*:321-331.

Savard M., C. Belanger, M. Tardif, P. Gourde, L. Flamand, and J. Gosselin. 2000. Infection of primary monocytes by Epstein-Barr virus. *J Virol.74*: 2612-2619.

Savard M., and J. Gosselin J. 2006. Epstein-Barr virus immunosuppression of innate immunity mediated by phagocytes. *Virus Research 119*:134-145.

Shannon-Low CD, Neuhierl B, Baldwin G, Rickinson AB, and Delecluse HJ. 2006. Resting B cells as a transfer vehicle for Epstein-Barr virus infection of epithelial cells. *Proc. Natl Acad Sci USA 103*:7065-7070.

Sharma V and Zhang, L. 2001. Interleukin-8 expression in AIDS-associated lymphoma B-cell lines. *Biochem Biophys Res Comm 282*:369-375.

Shimakage M, Kimura M, Yanoma S, Ibe M, Yokota S, Tsujino G, Kozuka T, Dezawa T, Tamura S, Ohshima A, Yutsudo M and Hakura, A. 1999. Expression of latent and replicative-infective genes of Epstein-Barr virus in macrophage. *Arch Virol. 144*:157-166.

Sommer P, Kremmer E, Bier S, Konig S, Zalud P, Zeppezauer M, Jones JF, Mueller-Lantzsch N and Grasser FA. 1996. Cloning and expression of the Epstein-Barr virus encoded dUTPase: Patients with acute, reactivated or chronic virus infection develop antibodies against the enzyme. *J. Gen. Virol* 77:2795-2805.

Stowe RP, Kozlova EV, Yetman DL, Walling DM, Goodwin JS and Glaser R. 2007. Chronic herpesvirus reactivation occurs in aging. *Exp Gerontol 42*:536-570.

Steinberg, D. 2002. Atherogenesis in perspective: hypercholesterolemia and inflammation as partners in crime. *Nature Med 8*: 1211-1217.

Steven NM, Annels NE, Kumar A, Leese AM, Kurilla MG and Rickinson AB. 1997. Immediate early and early lytic cycle proteins are frequent targets of the Epstein-Barr virus-induced cytotoxic T cell response. *J Exp Med. 185*:1605-1617.

Tanner JE, Alfieri C, Chatil TA and Diaz-Mitoma F. 1996. Induction of interleukin-6 after stimulation of human B-cell CD21 by Epstein-Barr virus glycoproteins gp350 and gp220. *J. Virol. 70*:570-575.

Tanner JE, Diaz-Mitoma F, Rooney CM, and AlWeri, C. 1997. Anti-interleukin-10 antibodies in patients with chronic active Epstein– Barr virus infection. *J. Infect. Dis. 176*: 1454–1461.

Tarbouriech N, Buisson M, Seigneurin JM, Cusack S and Burmeiste WP. 2005 The monomeric dUTPase from Epstein-Barr virus mimic trimeric dUTPases. *Structure 13*:1299-1310.

Thompson MP and Kurzrock R. 2004. Epstein-Barr virus and cancer. *Clin Cancer Res. 10*: 803-821.

Thorley-Lawson DA. 2001. Epstein-Barr virus: exploiting the immune system. *Nature Rev Immunol 1*: 75-82.

Touitou R, Arbach H, Cochet C, Feuillard J, Martin A, Raphael M and Joab I. 2003. Heterogeneous Epstein-Barr virus latent gene expression in AIDS-associated lymphomas and in type I Burkitt's lymphoma cell lines. *J Gen Virol. 84*:949-956.

Tugizov, S., Herrera, R., Veluppillai, P., Greenspan, J., Greenspan, D., Palefsky, J.M., 2007. Epstein-Barr Virus (EBV)-Infected Monocytes Facilitate Dissemination of EBV within the Oral Mucosal Epithelium. *Journal of Virology 81*, 5484-5496.

Turenne-Tessier M, Ooka T, Calander A and de The G, DJ. 1989. Relationship between nasopharyngeal carcinoma and high antibody titers to Epstein-Barr virus specific thymidine kinase. *Int J Cancer 43*: 45-48.

Vabulas, RM, Ahmad-Nejad P, daCosta C, Miethke T, Kirschning CJ, Hacker H and Wagner H. 2001. Endocytosed HSP60s use toll-like receptor 2 (TLR2) and TLR4 to activate the toll/interleukin-1 receptor signaling pathway in innate immune cells. *J. Biol. Chem. 275*:31332-31339.

Walling DM, Flaitz CM, Nichols CM, Hudnall SD and Adler-Storthz K. 2001. Persistent productive Epstein-Barr virus replication in normal epithelial cells in vivo. *J Inf. Dis. 184*:1499-1507.

Waldman W.J., M.V. Williams Jr., S. Lemeshow, P. Binkley, D. Guttridge, J.K. Kiecolt-Glaser, D.A. Knight, K.J. Ladner, and R. Glaser. 2008.Epstein-Barr virus-encoded dUTPase enhances proinflammatory cytokine production by macrophages in contact with endothelial cells: evidence for depression-induced atherosclerotic risk. *Brain Behavior Immunity. 22*:215-223.

Webster-Cyriaque J, Middledorp J and Raab-Traub N. 2000. Hairy leukoplakia: an unusual combination of transforming and permissive Epstein-Barr virus infections. *J Virol 74*:7610-7618.

Williams MV and Pollack JD. 1985. Pyrimidine deoxyribonucleotide metabolism in Mollicutes. *Int. J. Systemic Bacteriol. 35*: 227-230.

Williams, M.V., Holliday, J.E., Glaser, R., 1985. Induction of a deoxyuridine triphosphate nucleotidohydrolase activity in Epstein–Barr virus-infected cells. *Virology 142*,326–333.

Yang L, Aozasa K, Oshimi K and Takada K. 2004. Epstein-Barr virus (EBV)-encoded RNA promotes growth of EBV-infected T cells through interleukin-9 induction. *Cancer Res 64*:5332-5337.

Young LS and Rickinson AB. 2004. Epstein-Barr virus: 40 years on. Nature Rev. 4:757-768.

Zajac AJ, Blattman JN, Murali-Krishna K, Sourdive DJ, Suresh M, Altman JD and Ahmed R. 1998. Viral immune invasion due to persistence of activated T cells without effector function. *J Exp Med 188*:2205-2213.

Zuo J, Thomas W, van Leeuwen D, Middeldorp JM, Wiertz EJHJ, Ressing ME and Rowe M. 2008. The DNase of gamma-herpesviruses impairs recognition by virus-specific CD8+ T cells through an additional host shutoff function. *J Virol.* Published online ahead of print (Dec 19, 2007).

In: DNA Tumor Viruses
Editor: H. E. Tao
ISBN 978-1-60692-111-1

Chapter III

EBV VIRUS AND CANCER

Viroj Wiwanitkit
Wiwanitkit House, Bangkhae, Bangkok Thailand 10160.

ABSTRACT

Epstein Barr virus (EBV) is a DNA virus at can be the cause of several cancers. This DNA virus is proved as DNA tumor virus. Examples of EBV – related cancers include nasopharyngeal carcinoma and lymphoma. The reports on oncogenesis of EBV in significant cancers will be discussed in this chapter. Detection for the EBV is a commonly applied virological laboratory. In this work, the present novel knowledge on EBV laboratory diagnosis, therapy and prevention are reviewed and discussed. In addition, the author also performs a metanalysis on the previous reports on EBV virus infection in tropical countries.

INTRODUCTION TO EPSTEIN-BARR VIRUS

Epstein-Barr virus (EBV) is a well known virus in medicine. It is a large lymphotrophic DNA virus that establishes life-long residency in the infected host. EBV is related to several types of human cancer and immortalizes human B cells very efficiently [1]. EBV is a B-lymphotropic virus that is related to a range of human malignancies. EBV is a ubiquitous herpesvirus related to a variety of lymphoid and epithelial tumors [2]. In healthy lymphocytes and in tumors immune surveillance is evaded by suppression of a family of immunodominant viral antigens. Methylation of a viral promoter plays a significant function in this suppression [2]. Methylation of the viral genome in the latent state over evolutionary time is believed to account for CpG suppression that distinguishes this virus from most other large DNA viruses [2]. Methylation of cytosines within CpG dinucleotides at promoter regions is significant for gene silencing and genome integrity [3]. EBV has generated to use DNA methylation to maximize persistence and brings itself from immune detection [3]. EBV's reliance on DNA methylation also gives a unique therapeutic strategy for the therapy of EBV-associated tumors [3]. Chronic active EBV has been considered to be a non-neoplastic T-cell lymphoproliferative disease related to EBV infection [4]. In EBV-associated disorders, the cell phenotype-dependent differences in EBV latent gene expression may reflect the strategy of the virus in relation to latent infection [4]. Examples of EBV – related cancers include nasopharyngeal carcinoma and lymphoma. The reports on oncogenesis of EBV in significant cancers will be discussed in this chapter. Detection for the EBV is a commonly applied virological laboratory. In this work, the present novel knowledge on EBV laboratory diagnosis, therapy and prevention are reviewed and discussed. In addition, the author also performs a metanlysis on the previous reports on EBV virus infection in tropical countries.

Epstein-Barr Virus and Oncogenesis

As a common virus infection, EBV appears to have generated to exploit the process of B cell generation to persist as a life-long asymptomatic infection [5]. However, the virus can lead to oncogenesis as evidenced by its frequent detection in certain tumors. EBV maintains latent infection in cancer cells, and there are three types of latent infection (type I-III) according to the patterns of viral latent genes expression [6]. EBV has the ability to transform B cells into immortalized lymphoblastoid cell lines showing type III latency, in which all latent genes are expressed [6]. The mechanism of B-cell transformation has gived a model of EBV-associated lymphomas in immunosuppressed individuals. In type I and II latency, the limited numbers of latent genes are expressed [6]. Of interest, recent studies have clarified the unique pathogenic mechanism of this mysterious disease, and demonstrated the close relationship between human mosquito bite and EBV-carrying NK cell lymphocytosis; for example, CD4(+) T cells from the patients markedly responded to mosquito salivary gland extracts, and the CD4(+) T cells stimulated by mosquito bites may play a key function in the generation of HMB and NK cell oncogenesis via the induction of EBV reactivation and EBV-oncogene expression, respectively [7]. It is accepted that is EBV related to B-cell lymphomas in immunosuppressed patients as well as some cases of Burkitt's lymphoma, some T and natural killer lymphomas and approximately 40% of cases of Hodgkin's disease [8]. Human T-cell leukemia virus 1 and human herpes virus 8 genomes are also found in tumor cells in some types of lymphoma, while there are epidemiological information linking hepatitis C and lymphoma [8]. The presence of the viral genome in all these malignancies offers the prospect for therapeutic interventions targeting virus-encoded proteins [8]. Young and Murrey said that the transforming effects were related to the restricted expression of EBV genes such that only a subset of so-called latent virus proteins are expressed in virus infected tumors and in lymphoblastoid cell lines [5]. Young and Murrey also noted that distinct forms of EBV latency were manifest in the different tumors and these happened to be a vestige of the pattern of latent gene expression applied by the virus during the setting of persistent infection within the B cell pool [5].

EBV-encoded RNAs (EBERs) are the most abundant viral transcripts in latently EBV-infected cells [9]. Nanbo and Takada demonstrated that EBERs confer resistance to interferon (IFN)-alpha-induced apoptosis by inhibition of double-stranded (ds) RNA-activated protein kinase (PKR), which is the key mediator of the antiviral effect of IFN-alpha. These studies give a novel notion that RNA particles lead to oncogenesis [9]. Kitagawa et al suggested that interleukin-10 induced by EBERs acted as an autocrine growth factor for BL. EBERs, EBER1 and EBER2, are non-polyadenylated RNAs and are 166 and 172 nucleotides long, respectively [10]. These findings indicate that RNA particles could regulate cell growth [10]. It now appears that EBERs play a key function in maintaining the malignant phenotypes of Burkitt's lymphoma cells [11]. The EBERs confer clonability in soft agarose, tumorigenicity in mice, and resistance to apoptosis against various stimuli in Burkitt's lymphoma [11]. Furthermore, EBERs induce transcription of interleukin-10, which acts as an autocrine growth factor of Burkitt's lymphoma [11]. Ruf et al reported that wild-type or greater levels of EBER expression in EBV-negative cells did not promote Burkitt's lymphoma cell survival [12]. Ruf et al suggested that EBV could lead to Burkitt's lymphoma through at least two avenues: an EBER-dependent mechanism that enhances tumorigenic potential independent of a direct effect on apoptosis, and a second mechanism, mediated by an as-yet-unidentified EBV gene that offsets the proapoptotic consequences of deregulated c-MYC in Burkitt's lymphoma [12]. According to discovery of a binding site for the oncoprotein c-Myc at a central position of the EBV genome, Niller et al proposed for alternative pathogenetic pathways of EBV oncogenesis [13]. In the first scenario nuclear maintenance of the EBV genome and activation of viral anti-apoptotic functions with the help of c-Myc are indispensable for the origin of malignant tumors from the germinal centre B-cell [13]. In the second scenario expression of the main viral transforming protein EBNA2 is essential for immortalisation and non-malignant morphological transformation of any (germinal centre derived or non-germinal centre) B-cell in the absence of T cell control [13].

EBV AND LYMPHOMA

A. EBV and Lymphoma in General

Lymphoma is a common hematological malignancy. Relationship with the EBV was indicated [14]. Concerning the correlation between tumor and EBV, Busson *et al* recently said that tumor generation happened to require the expression of a small subset of transforming viral RNAs [15]. They also noted that Impairment of the interactions of viral proteins with cellular partners or disruption of viral latency might prove to be useful for novel therapeutic strategies [15]. Anti-EBV serology, immunoligcal test for detection of EBV infection is applied for screening and detection of several EBV – related cancers including lymphoma [16]. However, the detection of Anti-EBV IgG is not useful in some specific area such as the tropical Southeast Asia where most of the populations usually have positive titer for this antibody [17]. Therefore, novel more specific Anti-EBV tests are required. Of several tests, IgA antibody to viral capsid antigens of EBV (VCA-lgA antibody) has been mentioned as a valuable test for screening for and early detection of lymphoma [18-20].

Recently, Wiwanitkit performed an appraisal on the prevalence of serum VCA-lgA antibody among Thai patients with lymphoma comparing with healthy control subjects. Risk analysis was performed [21]. Wiwanitkit hypothesized that the serum VCA-lgA antibody might be an effective screening test for lymphoma among the Thais [21]. According to this work, a literature review to find the previous reports about prevalence of serum VCA-lgA antibody among patients with lymphoma and healthy control subjects in Thailand was performed. The author applied the electronic search engine PubMed (www.pubmed.com) in searching for the literatures. The author also reviewed the published works in all 256 local Thai journals, which is not included in the international citation index by the informationbase Thai Index Medicus. Any report that did not present the prevalence in both patients with lymphoma and healthy control subjects were excluded. The available reports were collected and extracted for the information about the prevalence of serum VCA-lgA antibody. Those primary information were applied for further metanalysis study. Concerning the metanalysis study, the summarized antibody positive rate in the patients and healthy subjects as well as odds ratio were calculated. For operative definition, the summarized antibody positive rate is equal to the summative number of positive case/summative number of all case. The odd ratio was calculated according to the general epidemiological study. According to the literature review, 2 reports [19-20] were recruited for further metanalysis. According to the metanalysis, 39 cases and 252 healthy subjects were investigated for serum VCA-lgA antibody [21]. The summarized antibody positive rate in the patients and healthy subjects are 35.9 % (14/39) and 2.4 % (6/252), respectively. The odds ratio is 22.96 [21]. In several tropical countries including Thailand, the reported prevalence of EBV infection is very high. There are some comments on the correlation between EBV and lymphoma in those countries [19,20]. According to this study, there are some efforts to use EBV serological markers for lymphoma in Thailand [19,20]. Wiwanitkit tried to summarize the previous reported on the prevalence of serum VCA-lgA antibody among Thai patients with lymphoma comparing with healthy control subjects. According to this study, it could be seen that having serum VCA-lgA antibody positive is a very high risk for lymphoma.

B. EBV and Burkitt's Lymphoma

Subsequent studies have proven that EBV is the causative agent in most cases of infectious mononucleosis. Primary infection is usually asymptomatic in childhood; but in adulthood, it is related to a self-limiting infectious mononucleosis syndrome in approximately one third of the cases [22]. The EBV nuclear antigen 1 (EBNA1) protein can support the synthesis and maintenance of the viral genome [23]. Burkitt's lymphoma, a tumor occurring in endemic, sporadic and AIDS-associated forms, is the classic example of a human malignancy whose pathogenesis involves a specific cellular genetic change, namely, a chromosomal translocation deregulating expression of the c-myc oncogene, complemented in several cases by the action of an oncogenic virus, the EBV [24]. It has been known for several years that the fundamental transforming event in Burkitt's lymphoma is the translocation of the MYC gene, and the events that bring about this translocation and those that allow cells to survive with the constitutive expression of MYC have been the subject of intense investigation [25]. Novel information show that inhibiting EBNA1 in Burkitt's lymphoma cells induces cell death by apoptosis [23]. Therefore, EBNA1 inhibits apoptosis and, according to recent findings, does so independently of other viral genes. The latent membrane protein 2a (LMP2a) binds to signaling particles that are engaged by the B-cell receptor and inhibits the signaling that is mediated by antigen binding [23]. However, several mechanistic processes have been unravelled in EBV-associated tumors whereby the virus may modify the cellular phenotype and may influence the interaction of tumor cells with their microenvironment [26]. Takada suggested that EBV infection would upregulate expression of bcl-2 protein to protect cells from c-myc-induced apoptosis, and to allow c-myc to exert its oncogenic functions . The virus carrier state can lead to the evasion of apoptosis and can intensify the response to growth promoting signals, too [26]. In 2004, van den Bosch proposed that endemic Burkitt's lymphoma was an alliance between three infections and a tumor promoter [28]. van den Bosch reported that extracts of *Euphorbia tirucalli* were tumor promoters and could induce the characteristic 8;14 translocation of endemic Burkitt's lymphoma in EBV-infected cell-lines [28]. van den Bosch mentioned that they also activate latent EBV in infected cells, enhance EBV-mediated cell transformation, and modulate EBV-specific immunity [28].

In the novel WHO classification, the category of Burkitt's lymphoma includes classic Burkitt's lymphoma and a variant-Burkitt-like-lymphoma. In addition, three subcategories; endemic, non-endemic, and immunodeficiency-associated were proposed to reflect the major clinical and genetic subtypes of this disease. Endemic Burkitt's lymphoma is well known to carry EBV. However, not more than 20% of the sporadic Burkitt's lymphoma carry EBV [29].

C. EBV and Lymphoma in Human Immunodeficiency Virus Infection

For correlation between EBV and lymphomagenesis in human immunodeficiency virus (HIV(infected patients, three main factors are predominant in HIV-related lymphomagenesis: cellular immunodeficiency, oncogene viruses (EBV) and molecular pathological lesions [30]. Basically, the pathogenesis of HIV-related lymphoma is a multistep process involving factors gived by the host, as well as alterations intrinsic to the tumor one [31]. Host factors involved in HIV-related lymphomagenesis include reduced immunosurveillance particularly against EBV-infected B cells, HIV-induced alteration of endothelial functions, B-cell stimulation and selection by antigen, HIV-induced deregulation of several cytokine loops, and possibly the host's genetic background [31]. HIV-related cellular immunodeficiency leads to the increase of EBV infected B-cells and to the diminution of antitumor immunity [30,32]. Indeed, clonal EBV genome is found in lymphoma cells in 30 to 70% of cases of HIV-related NHL [30]. It expresses oncogenic proteins including LMP-1 which behaves like an activated CD40 and it also induces the expression of intra-cellular genes which stimulate cell growth and inhibit apoptosis [30]. Recently, Fais et al performed an analysis of stepwise genetic changes in an AIDS-related BL [33]. According to this work, it is proposed that stimulation by an antigen or a superantigen initially favored the clonal expansion and accumulation of other cytogenetic changes, including those involved in receptor editing [33]. These events occurred prior to or during the germinal center (GC) phase of B-cell maturation [33]. Thereafter, possibly upon exit of the cells from the GC, EBV infection occurred, further promoting lymphomagenesis [33]. Concerning the type of EBV in Hodgkin's disease in HIV – infected patients, Santon and Bellas found that type 2 EBV was in 26.7% of HIV-associated Hodgkin's disease while the existence of double infections by type 1 and 2 EBV was also observed in 6.7% of HIV-associated Hodgkin's disease [34]. In addition, they found that the latent membrane protein-1 (LMP-1) 30-base pair (bp) deletions in the Hodgkin's disease was identified in 83.3% of HIV-positive cases [34]. There have been limited and controversial information on the effect of antiretroviral drug use on the natural history of EBV-related lymphoma.

EBV and Nasopharyngeal Carcinoma

Incidence of nasopharyngeal carcinoma has remained high in endemic regions [35]. Diagnosing the disease in the early stages requires a high index of clinical acumen and, although most cross-sectional imaging investigations show the tumor with precision, confirmation is dependent on histology [36]. This cancer is rare in occidental countries (<1/100,000) and relatively moderate to highly frequent in the Mediterranean area and south-east Asia [37]. Strong correlation between EBV and nasopharyngeal carcinoma is proposed [36]. This disease is related to Epstein-Barr virus with a latent infection starting in the oropharyngeal epithelium and involving not only both epithelial tumor cells but also B lymphocytes [37]. EBV - encoded RNA signal is present in all nasopharyngeal carcinoma cells, and early diagnosis of the disease is possible through the detection of raised antibodies against EBV [35]. Anti-EBV serology is suggestive of for NPC with an elevated level of IgA EA (early antigen) and VCA (viral capsid antigen) [37]. Cyfra 21 represents a promising serum marker for nasopharyngeal with a 80% sensitivity. EBV-encoded RNA hybridization and EBV LMP1 immunostains are also applied routinely to detect latent EBV in tissues affected by posttransplant lymphoproliferative disorder (PTLD) or in enlarged nodes from patients with infectious mononucleosis [38]. Traditional serology is the best test for evaluating acute versus remote infection in healthy individuals [38]. High serological titers serve as a tumor marker for some EBV-related malignancies, but titers are not a dependable tumor marker in immunocompromised hosts. EBV viral load testing by quantitative DNA amplification of blood samples is a promising novel laboratory test that has proven useful for early diagnosis and monitoring patients with PTLD [38].

Radiotherapy remains the base of loco-regional therapy with a more frequent and systematic use of systemic chemotherapy (primary or concomitant) for high-risk-patients (T3-4 and N2-3 disease) [37]. Chan et al reviewed the current evidence available on the various chemotherapy-radiotherapy sequencing approaches and seek to define the optimal integration of radiotherapy and chemotherapy [39]. Despite consistently high response rates to platinum-based neoadjuvant chemotherapy, none of six randomized studies of neoadjuvant and/or adjuvant chemotherapy presented any improvement in summarized survival, although two did demonstrate significant improvement in local control rates and progression-free survival. However, three randomized studies of concurrent cisplatin-radiotherapy one with, and two without, adjuvant chemotherapy demonstrated significant improvement in progression-free survival and two of these have demonstrated improvement in summarized survival [39]. Chan et al concluded that preliminary information on the use of neoadjuvant chemotherapy followed by concurrent chemoradiation had been highly encouraging [39].

Table 1. Some interesting report on EBV and nasopharyngeal carcinoma from Thailand

Authors	Details
Puthavathana et al [40]	This study reported on the prevalence of specific IgA and IgG antibodies to EBV viral capsid antigen in nasopharyngeal carcinoma patients with different histological types of carcinoma and their age-matched controls by the indirect immunofluorescence test, using the B-95-8 lymphoblastoid cell line as source of viral capsid antigen [40]. According to this work, EBV IgA was found in all histological types of nasopharyngeal carcinoma [40].Puthavathana et al said that use of anti-EBV IgA for monitoring cancer therapy was to be further investigated [40].
Neimhom et al [41]	Niemhom et al investigated EBV-DNA in all the three types of nasopharyngeal carcinoma to assess the association of EBV and nasopharyngeal carcinoma in Thai patients. [41] EBV-DNA was determined in formalin-fixed paraffin-embedded tissues using polymerase chain interrelation [41]. According to this work, EBV-DNA was demonstrated in most cases of squamous cell NPC but the association of EBV in this type of carcinoma was not as frequent as in the other two types of nasopharyngeal carcinoma [41].
Luxamechanporn et al [42]	This study applied the PCR technique as a screening test for early detection of EBV DNA in this group and also attempted to compare the outcome with that of IgA antibody level [42]. All specimens studied were negative for EBV DNA and low IgA antibody titer obtained [42]. The negative outcomes implied that either the sample studied was inadequate to detect the low percentage of positivity of EBV in a high risk population or prolonged infection of EBV in nasopharyngeal mucosal cells is not necessary for initiation of nasopharyngeal carcinoma [42].

EBV AND GASTRIC CARCINOMA

Gastric carcinoma is another cancer that is commonly mentioned for the correlation to EBV infection. The EBV is determined in the tissue of about 10% of gastric carcinoma cases throughout the world [43]. In each case, 100% of carcinoma cells are infected with EBV [43]. Analysis of EBV in carcinoma biopsies indicates that carcinoma is formed by the proliferation of a single EBV infected cell [43]. These findings suggest that EBV plays an significant function in the generation of EBV positive gastric carcinomas [43]. In order to study the mechanism of Epstein-Barr virus (EBV) infection in gastric carcinoma cells, Luo et al characterized the EBV infection in signet ring cell line HSC-39, derived from a human gastric carcinoma [44]. Luo et al demonstrated that the cellular phenotypes of HSC-39 cells were altered by EBV infection in strain-specific manner [44]. Luo et al proposed the HSC-39 cell line as a model target for the study of the mechanism and significance of EBV infection in gastric carcinoma [44]. In 1999, Sairenji characterized two EBV positive cell lines, GT38 and GT39 from gastric tissues [45]. Sairenji found that the both cell lines were EBV latency type III and produced the virus spontaneously and the character of tumor cells was demonstrated by the colony formation in soft agar and the tumorigenesis in severe combined immunodeficient mice [45]. In 2004, Wang et al performed a study to determine the expression of EBV latent and lytic infection genes in gastric carcinoma at the transcriptional level by RT-PCR and Southern hybridization,and investigate the relationship between EBV-encoded genes and the tumorigenesis of gastric carcinoma at the molecular level [46]. According to this work, the latent pattern of EBV in EBVaGCs corresponds to the latency I or unique latency I/II, intermediate between the latency I and II. Part of lytic infection genes were expressed in EBVaGCs tissues [46]. Wang et al concluded that BARF1 and BHRF1 genes expressed in part of gastric carcinomas and their functions in gastric tumorigenesis needed to be further studied [46]. Shousha and Luqmani said that lymphoepithelioma-like gastric carcinoma seemed to be the main, if not the only, EBV positive gastric cancer [47]. Shousha and Luqmani noted that the presence of EBV positive normal gastric and duodenal cells suggested that these cells may acted as a reservoir for the virus in some people--a possibility that deserves wider investigation [47]. Of interest, Luqmani et al reported that EBV infection occurred only in the lymphoepithelioma type of gastric carcinoma and was absent from the other adenocarcinomas and from normal and benign tissues [48]. Luqmani et al also said that the occasional EBER positive interrelation encountered in normal cells was probably the outcome of a false signal arising from neuroendocrine cells as a consequence of the biotin-containing detection system [48].

EBV and other Carcinomas

EBV is an etiological agent in infectious mononucleosis, is implicated in some malignant lymphomas and lymphoepithelioma-like carcinomas, and has been sporadically reported in carcinomas of the breast, lung, and other sites [49]. Concerning breast malignancy, recent information give substantial, but not conclusive, evidence that human papillomavirus, the mouse mammary tumor virus and Epstein-Barr virus may have a function in the etiology of human breast cancer [50]. Lymphoepithelioma-like carcinomas (LELCs) are tumors with morphologic features identical to those of undifferentiated nasopharyngeal carcinoma. They are characterized by a pronounced lymphocytic infiltrate, often obscuring the neoplastic epithelial component [51]. Less than 50 cases of LELCs are reported in the literature. This type of breast cancer is mentioned for the possible correlation to EBV infection. However, there is no exact report [51-55]. Although the prognosis of LELC-B is thought to be favorable, which is also similar to medullary carcinoma of the breast (MC-B), distant metastasis was determined in the present case [52]. To confirm the clinicopathological entity of these two tumors, it is significant to recognize the difference between LELC-B and MC-B [52].

EBV replication within type II alveolar cells was shown to occur in adult cryptogenic fibrosing alveolitis [56]. Latent membrane protein 1 is one of the Epstein-Barr-virus-associated proteins and is expressed on the surface of EBV-infected cells in the latent and replicating phases [56]. Latent membrane protein 1 was positive in the cuboidal epithelial cells of the lungs from some patients with idiopathic pulmonary fibrosis, and that its positivity correlated with poor prognosis [53]. However, there was also lack of evidence for a function of Epstein-Barr virus in the increase of lung cancer in idiopathic pulmonary fibrosis [56]. For lung cancer, LELC of the lung was first reported in 1987. Histologically, it is indistinguishable from undifferentiated nasopharyngeal carcinoma. The carcinogenic function of latent EBV infection in causing LELC of the lung has been evident almost exclusively in Asians compared with Caucasians [57,58]. Among the reported cases, more than half were in early resectable stages (I or II) and there was a trend for peribronchovascular spread with vascular encasement in advanced disorders [57]. The summarized survival is more favourable in LELC of the lung compared with non-LELC type of non-small cell lung carcinoma [57].

REFERENCES

[1] Farrell PJ. Epstein-Barr virus immortalizing genes. *Trends Microbiol.* 1995 Mar;3(3):105-9.

[2] Ambinder RF, Robertson KD, Tao Q. DNA methylation and the Epstein-Barr virus. *Semin Cancer Biol.* 1999 Oct;9(5):369-75.

[3] Tao Q, Robertson KD. Stealth technology: how Epstein-Barr virus uses DNA methylation to cloak itself from immune detection. *Clin Immunol.* 2003 Oct;109(1):53-63.

[4] Yoshioka M, Kikuta H, Ishiguro N, Ma X, Kobayashi K. Unique Epstein-Barr virus (EBV) latent gene expression, EBNA promoter usage and EBNA promoter methylation status in chronic active EBV infection. *J Gen Virol.* 2003 May;84(Pt 5):1133-40.

[5] Young LS, Murray PG. Epstein-Barr virus and oncogenesis: from latent genes to tumors. *Oncogene.* 2003 Aug 11;22(33):5108-21.

[6] Iwakiri D, Samanta M, Takada K. Mechanistic processes of EBV-mediated oncogenesis. *Uirusu.* 2006 Dec;56(2):201-8

[7] Asada H. Hypersensitivity to mosquito bites: a unique pathogenic mechanism linking Epstein-Barr virus infection, allergy and oncogenesis. *J Dermatol Sci.* 2007 Mar;45(3):153-60.

[8] Ahmed N, Heslop HE. Viral lymphomagenesis. *Curr Opin Hematol.* 2006 Jul;13(4):254-9.

[9] Nanbo A, Takada K. The function of Epstein-Barr virus-encoded small RNAs (EBERs) in oncogenesis. *Rev Med Virol.* 2002 Sep-Oct;12(5):321-6.

[10] Kitagawa N, Goto M, Kurozumi K, Maruo S, Fukayama M, Naoe T, Yasukawa M, Hino K, Suzuki T, Todo S, Takada K. Epstein-Barr virus-encoded poly(A)(-) RNA supports Burkitt's lymphoma growth through interleukin-10 induction. *EMBO J.* 2000 Dec 15;19(24):6742-50.

[11]Takada K, Nanbo A. The function of EBERs in oncogenesis. *Semin Cancer Biol.* 2001 Dec;11(6):461-7.

[12]Ruf IK, Rhyne PW, Yang C, Cleveland JL, Sample JT. Epstein-Barr virus small RNAs potentiate tumorigenicity of Burkitt lymphoma cells independently of an effect on apoptosis. *J Virol.* 2000 Nov;74(21):10223-8.

[13]Niller HH, Salamon D, Ilg K, Koroknai A, Banati F, Schwarzmann F, Wolf H, Minarovits J. EBV-associated neoplasms: alternative pathogenetic pathways. *Med Hypotheses*. 2004;62(3):387-91.

[14]Young LS, Rickinson AB. Epstein-Barr virus: 40 years on. *Nat Rev Cancer* 2004;4:757-68

[15]Boussen H, Bouaouina N, Mokni-Baizig N, Gamoudi A, Chouchane L, Benna F, Ladgham A. Nasopharyngeal carcinoma. Recent information. *Pathol Biol (Paris)* 2005;53:45-51

[16]Liang R. Diagnosis and management of primary nasal lymphoma of T-cell or NK-cell origin. *Clin Lymphoma* 2000;1:33-7

[17]Pancharoen C, Mekmullica J, Chinratanapisit S, Bhattarakosol P, Thisyakorn U. Seroprevalence of Epstein-Barr virus antibody among children in various age groups in Bangkok, Thailand. *Asian Pac J Allergy Immunol* 2001;19:135-7

[18]Trzcinska A, Litwinska B. Association of EBV infection with lymphomas. *Acta Microbiol Pol* 2003;52:191-4

[19]Srivatanakul P, Tiwawech D, Boonyaratavej C, Sombooncharoen S. Early detection of nasopharyngeal carcinoma by using serum IgA antibody to Epstein Barr virus as a screening test *Thai Cancer J* 1988; 14: 18-24

[20]Srivatanakul P, Sukvirach S, Puribhat S, Boonyaratavej C. Significance of IgA and IgG antibodies to epstein-barr virus in Thai patients with nasopharyngeal carcinoma and lymphoma *Thai Cancer J* 1981; 7: 21-25

[21]Wiwanitkit V. Summary on the prevalence of serum VCA-IgA antibody among Thai patients with lymphoma: an Implication for risk and screening purpose. *Int J Trop Med* 2005; 2(2): 1.

[22]Rezk SA, Weiss LM. Epstein-Barr virus-associated lymphoproliferative disorders. *Hum Pathol.* 2007 Sep;38(9):1293-304.

[23]Hammerschmidt W, Sugden B. Epstein-Barr virus sustains Burkitt's lymphomas and Hodgkin's disease. *Trends Mol Med.* 2004 Jul;10(7):331-6.

[24]Kelly GL, Rickinson AB. Burkitt lymphoma: revisiting the pathogenesis of a virus-associated malignancy. *Hematology Am Soc Hematol Educ Program.* 2007;2007:277-84.

[25]Brady G, MacArthur GJ, Farrell PJ. Epstein-Barr virus and Burkitt lymphoma. *J Clin Pathol.* 2007 Dec;60(12):1397-402.

[26]Klein E, Kis LL, Takahara M. Pathogenesis of Epstein-Barr virus (EBV)-carrying lymphomas. *Acta Microbiol Immunol Hung.* 2006 Dec;53(4):441-57.

[27]Takada K. Function of Epstein-Barr virus in Burkitt's lymphoma. *Curr Top Microbiol Immunol.* 2001;258:141-51.

[28]van den Bosch CA. Is endemic Burkitt's lymphoma an alliance between three infections and a tumor promoter? *Lancet Oncol.* 2004 Dec;5(12):738-46.

[29]Ohnishi K. Clinicopathological characteristics of Burkitt lymphoma. *Nippon Rinsho.* 2000 Mar;58(3):635-8.

[30]Takada K, Nanbo A. The function of EBERs in oncogenesis. *Semin Cancer Biol.* 2001 Dec;11(6):461-7.

[31]Besson C, Raphael M. Lymphoma genesis in the context of HIV infection. *Ann Med Interne (Paris).* 2003 Dec;154(8):523-8.

[32]Gaidano G, Capello D, Carbone A. The molecular basis of acquired immunodeficiency syndrome-related lymphomagenesis. *Semin Oncol.* 2000 Aug;27(4):431-41.

[33]Fais F, Fronza G, Roncella S, Inga A, Campomenosi P, Cutrona G, Pezzolo A, Fedeli F, Abbondandolo A, Chiorazzi N, Pistoia V, Ferrarini M. Analysis of stepwise genetic changes in an AIDS-related Burkitt's lymphoma. *Int J Cancer.* 2000 Dec 1;88(5):744-50.

[34]Santon A, Bellas C. Deletions within the epstein-barr virus latent membrane protein-1 oncogene in adult ordinary, HIV-associated and paediatric Hodgkin's disease. *Leuk Lymphoma.* 2001 Jan;40(3-4):235-42.

[35]Wei WI, Sham JS. Nasopharyngeal carcinoma. *Lancet.* 2005 Jun 11-17;365(9476):2041-54.

[36]Wakisaka N, Furukawa M. EBV and nasopharyngeal carcinoma. *Nippon Rinsho.* 2006 Mar;64 Suppl 3:647-50.

[37]Boussen H, Bouaouina N, Mokni-Baizig N, Gamoudi A, Chouchane L, Benna F, Ladgham A. Nasopharyngeal carcinoma. Recent information. *Pathol Biol (Paris).* 2005 Feb;53(1):45-51.

[38]Gulley ML. Molecular diagnosis of Epstein-Barr virus-related disorders. *J Mol Diagn.* 2001 Feb;3(1):1-10.

[39]Chan AT, Teo PM, Huang DP. Pathogenesis and therapy of nasopharyngeal carcinoma. *Semin Oncol.* 2004 Dec;31(6):794-801

[40]Puthavathana P, Sutthent R, Vitavasiri A, Wasi C, Chantarakul N, Thongcharoen P. Epstein-Barr virus serological markers for nasopharyngeal carcinoma in Thailand. *Southeast Asian J Trop Med Public Health.* 1991 Sep;22(3):326-31.

[41]Niemhom S, Maeda S, Raksakait K, Petchclai B. Epstein-Barr virus DNA in nasopharyngeal carcinoma in Thai patients at Ramathibodi Hospital, Bangkok. *Southeast Asian J Trop Med Public Health.* 1995;26 Suppl 1:325-8.

[42]Luxamechanporn T, Neimhom S, Kunachak S, Kulapaditharom B. Polymerase chain interrelation study of Epstein-Barr virus in high risk group of nasopharyngeal carcinoma: a preliminary report. *J Med Assoc Thai.* 1999 Jun;82(6):593-7.

[43]Takada K. Epstein-Barr virus and gastric carcinoma. ol Pathol. 2000 Oct;53(5):255-61.

[44]Luo B, Murakami M, Fukuda M, Fujioka A, Yanagihara K, Sairenji T. Characterization of Epstein-Barr virus infection in a human signet ring cell gastric carcinoma cell line, HSC-39. *Microbes Infect.* 2004 Apr;6(5):429-39.

[45]Sairenji T. Epstein-Barr virus (EBV) infection and gastric carcinoma: the approach through EBV infected epithelial cell lines. *Jpn J Infect Dis.* 1999 Jun;52(3):110-2.

[46]Wang Y, Luo B, Zhao P, Huang BH. Expression of Epstein-Barr virus genes in EBV-associated gastric carcinoma. *Ai Zheng.* 2004 Jul;23(7):782-7.

[47]Shousha S, Luqmani YA. Epstein-Barr virus in gastric carcinoma and adjacent normal gastric and duodenal mucosa. *Clin Pathol.* 1994 Aug;47(8):695-8.

[48]Luqmani YA, Linjawi SO, Shousha S. Detection of Epstein-Barr virus in gastrectomy specimens. *Oncol Rep*. 2001 Sep-Oct;8(5):995-9.

[49]Grinstein S, Preciado MV, Gattuso P, Chabay PA, Warren WH, De Matteo E, Gould VE. Demonstration of Epstein-Barr virus in carcinomas of various sites. *Cancer Res*. 2002 Sep 1;62(17):4876-8.

[50]Lawson JS, Günzburg WH, Whitaker NJ. Viruses and human breast cancer. *Future Microbiol*. 2006 Jun;1:33-51.

[51]Ilvan S, Celik V, Ulker Akyildiz E, Senel Bese N, Ramazanoglu R, Calay Z. Lymphoepithelioma-like carcinoma of the breast: is it a distinct entity? Clinicopathological evaluation of two cases and review of the literature. *Breast*. 2004 Dec;13(6):522-6.

[52]Kurose A, Ichinohasama R, Kanno H, Kobayashi T, Ishida M, Nishinari N, Sawai T. Lymphoepithelioma-like carcinoma of the breast. Report of a case with the first electron microscopic study and review of the literature. *Virchows Arch*. 2005 Sep;447(3):653-9.

[53]Sanati S, Ayala AG, Middleton LP. Lymphoepithelioma-like carcinoma of the breast: report of a case mimicking lymphoma. *Ann Diagn Pathol*. 2004 Oct;8(5):309-15.

[54]Pagano JS, Blaser M, Buendia MA, Damania B, Khalili K, Raab-Traub N, Roizman B. Infectious agents and cancer: criteria for a causal relation. *Semin Cancer Biol*. 2004 Dec;14(6):453-71.

[55]Mant C, Hodgson S, Hobday R, D'Arrigo C, Cason J. A viral aetiology for breast cancer: time to re-examine the postulate. *Intervirology*. 2004;47(1):2-13.

[56]Marzouk K, Corate L, Saleh S, Sharma OP. Epstein-Barr-virus-induced interstitial lung disease. *Curr Opin Pulm Med*. 2005 Sep;11(5):456-60.

[57]Ho JC, Wong MP, Lam WK. Lymphoepithelioma-like carcinoma of the lung. *Respirology*. 2006 Sep;11(5):539-45

[58]Castro CY, Ostrowski ML, Barrios R, Green LK, Popper HH, Powell S, Cagle PT, Ro JY. Relationship between Epstein-Barr virus and lymphoepithelioma-like carcinoma of the lung: a clinicopathologic study of 6 cases and review of the literature. *Hum Pathol*. 2001 Aug;32(8):863-72.

In: DNA Tumor Viruses
Editor: H. E. Tao

ISBN 978-1-60692-111-1

Chapter IV

DNA TUMOR VIRUSES: ONCOGENESIS OF HUMAN PAPILLOMAVIRUS (HPV)

Masachika Senba[1], Naoki Mori[2] and Akihiro Wada[3]

[1]Department of Pathology, Institute of Tropical Medicine, Nagasaki University, Nagasaki 852-5152, Japan;

[2]Division of Molecular Virology and Oncology, Gradual School of Medicine, University of the Ryukyus, Okinawa 903-0215, Japan;

[3]Department of Microbiology, Institute of Tropical Medicine, Nagasaki University, Nagasaki 852-8152, Japan.

ABSTRACT

Human papillomaviruses (HPV) have been etiologically associated with malignant lesions, most notably cervical, penile, anal and oro-pharynxal. Approximately 40 HPV types have been associated with lesions of the anogenital tract and these can be further classified into so-called low-risk and high-risk types based on their association with clinical lesions. While low-risk HPVs are generally associated with benign lesions such as condyloma acuminatum, infections with high-risk HPVs are associated with a significantly increased risk of developing cancer. Based on epidemiological and moleclular biological evidence, it is widely accepted that high risk HPVs, including HPV types 16 and 18, play an etiological role in the development of cancer. Two oncogenic viral proteins, E6 and E7, involved in cell cycle control, are required for the efficient immortalization of their natural host cells, resulting in primary human squamous epithelial cells developing mucosal and skin cancers. These proteins have been the focus of many studies of cervical cancer during the past 20 years, and their oncogenic mechanisms are now established E7 binds to and degrades the tumor suppressor Rb, while E6 binds to and inactivates the tumor suppressor p53. The causal relationship between chronic inflammation and cancer is widely accepted. Specifically, there is a strong association between tumor viruses and development of cancers. Activation of NF-kB is also involved cancer development and progression. Therefore, this short review focuses on the molecular players during the development from chronic inflammation to cancer.

INTRODUCTION

Histopathological characteristics of chronic inflammation include the predominance of lymphocytes and macrophages. In 1863, Virchow hypothesized that malignant tumors occurred in the chronic inflammation lesions associated with lymphocyte infiltration (Balkwill, & Mantovani, 2001; Coussens, & Werb, 2002; Lu, Ouyang, & Huang, 2006; Schottenfeld & Beebe-Dimmer, 2006; Hussan, & Harris, 2007). Activated lymphocytes and macrophages are interactive in releasing inflammatory mediators such as cytokines, chemokines, growth factors, and survival factors interact with specific cell surface receptors resulting in amplified immune reactivity.

The mechanisms of oncogenesis associated with infection and inflammation have not been elucidated. However, many oncogenic mechanisms have been proposed for infection and inflammation; for example, 1) Signal transduction: Toll-like receptors (TLR) for proinflammatory cytokins, such as tumor necrosis factor-α (TNF-α) and interleukin-1 (IL-1), IL-6, IL-8, and IL-18 leads to activation of many important signaling pathways. The other pro-inflammatory gene products are matrix metalloproteinase-9 (MMP-9), vascular endothelial growth factor (VEGF), cyclooxygenase-2 (COX-2), and 5-lipoxygenase (5-LOX). It is well accepted that the NF-κB group of transcription factors play a still more important role in inflammation. 2) Direct action of viral proteins and host cells: Products of binding viral proteins and human genes lead to alteration of host cells. Tumor-suppressor gene products of retinoblastoma protein (pRb) and p53 are inactivated by interaction with products of viral proteins. For example, human papillomavirus (HPV) E6 and E7 proteins are oncogenic factors in HPV related cancers. 3) Immunosuppression: Human immunodeficiency virus (HIV) causes immunosuppression, which leads to Kaposi's sarcoma and/or malignant lymphomas. 4) Production of reactive oxygen species (ROS) and reactive nitrogen species (RNS). Prolonged activation of inflammatory cells generates ROS and RNS that can damage host DNA, which can contribute to carcinogenesis (Hanahan, & Weinberg, 2000; Ohshima, Tatemichi, & Sawa, 2003; Karin, 2006; Aggarwal, Shishodia, Sandur, Pandey, & Sethi, 2006; Lu, Ouyang, & Huang, 2006).

The proportion of total cancer deaths attributable to infectious agents is estimated to be 20% to 25% in developing countries and 7% to 10% in industrialized countries (Pisani, Parkin, Munoz, & Ferlay, 1997). A causal relationship between chronic inflammation and cancer is widely accepted. Specifically, there is a strong association between tumor viruses and the development of human cancers. Both DNA and RNA viruses have been reported to be capable of causing cancer. Epstein-Barr virus (EBV), HPV, hepatitis B virus (HBV), and Kaposi's sarcoma associated herpesvirus (KSHV) are DNA viruses known to lead to the development of cancers. Cancer inducing RNA viruses include Hepatitis C virus (HCV) and human T lymphotropic virus type-1 (HTLV-1) (Mueller, 1995; Hausen, 1999; 2001; Damania, 2007). EBV is associated with Burkitt's lymphoma, nasopharyngeal carcinoma,

Hodgkin's lymphoma and non-Hodgkin's lymphoma; HBV and HCV viruses with hepatocellular carcinoma; HTLV-1 with adult T-cell leukemia/lymphoma; and HPV with cervical cancer, penile cancer, anal cancer, oropharyngeal cancer, and skin cancer (Table 1). In developing countries, 15% to 20% of the global cancer burden can be linked to these infections. An additional 5% can be attributed to bacterial infection (*Helicobactor pyrori*) and parasitic infections (*Schistosoma, Opistorchis, Clonorchis*) (Schottenfeld, & Beebe-Dimmer, 2006). Although many molecular investigators have been proposed oncogenesis due to inflammation associated cancer development, the mechanisms underlying the relationship between chronic inflammation and cancer still remain unresolved.

This review focuses on the potential role of HPV in the development of cancers, highlighting the recent advances in the understanding of the molecular mechanism by which HPV-encoded proteins may contribute towards tumor formation. Furthermore, the mechanism of E6 and E7 induced multiple step carcinogenesis and current identified molecular functions of relationship between infection and cancer are reviewed.

Table 1. Cronic inflammation and cancer

Microbe	Malignancy
Hepatitis B virus (HBV)	Hepatocelluar carcinoma
Hepatitis C virus (HCV)	Hepatocelluar carcinoma
Epstein-Barr virus (EBV)	Burkitt's lymphoma
	Hodgkin's lymphoma
	Nasopharingeal carcinoma
Human papillomavirus (HPV)	Cervical cancer
	Penile cancer
	Anal cancer
	Oropharyngeal carcinoma
	Skin cancer
Human T-lymphotropic virus type 1 (HTLV-1)	Adult T-cell leukemia/lymphoma
Human immunodeficiency virus (HIV) and	Kaposi's sarcoma
Kaposi's sarcoma associated herpesvirus (KSHV)	Non-Hodgkin' lymphoma
	Multicentric Castleman's disease
Helicobacter pylori	Adenocarcinoma of stomach
Schistosoma hematobium	Bladder carcinoma
Opisthorchis viverrini	Biliary cancer

HPV Infection and Cancers

Between 2000 and 2004, cancer of the cervical uteri was the 13th most common cancer among women in the United States, with an incidence rate of 8.8 per 100,000 (Espey, 2007). Cervical cancer is the second most common cancer among women worldwide, with an estimated 493,000 new cases and 274,000 deaths in 2002 (Parkin, & Bray, 2006). 83% of world-wide cases of this cancer occur in developing countries, such as sub-Sahara Africa, Melanesia, Latin America and the Caribbean, South-Central Asia, and Southeast Asia (Parkin, 2001; Parkin, & Bray, 2006). While the causative relationship between HPV infection and squamous cell cancers (SCC) of the genital tract is well established (Bosch, et al., 1995; Walboomers, et al., 1999; & Bryan, et al., 2006), the role of HPV in the development of malignancy is not yet clear. However, there is increasing evidence showing the involvement of HPV types in the process of mucosal and skin carcinogenesis. HPV infection is the most common sexually transmitted disease, thus, the risk factor for HPV infection is increased by certain sexual behaviors (Munoz, 2000; Bekkers, Massuger, Bulten, & Melchers, 2004). Cervical cancer is the most common cancer of woman in developing countries. The most common oncogenic HPV genotypes causing cervical cancer are 16 and 18. HPV-16 is the most prevalent high-risk HPV type and is found in cervical cancer cases and HPV-18 is the second most prevalent high-risk HPV. HPV are believed to be the primary causal agents for development of benign and malignant tumors in mucosal and skin lesions, and high-risk types such as type 16 and 18 are associated with more than 90% of all cervical cancer. HPV-6 and HPV-11 are usually associated with common genital condyloma. Bosch et al., (1995) reported that HPV-16 infection is predominately observed in squamous cell carcinoma whereas HPV-18 is predominately observed in adenocarcinoma. The majority cases of cervical cancer are squamous cell carcinoma, and adenocarcinoma is less common.

Penile cancer is rare disease in Europe and United States with incidence rates varying from 0.5 to 1.5 per 100,000 men. In developing countries, it is a common cancer with an incidence of 2 to 5 per 100,000 men (Tornesello, et al., 2008). The prevalence of HPV observed in penile cancer ranges from 30% to 80%, depending on the sensitivity of the detection methods and the selection of tumor type (McCance, et al., 1986; Varma, et al., 1991; Sarkar, Miles, Plieth, & Crissman, 1992; Scinicariello, & Rady, P., Saltzstein, Orihuela, & Tyring, 1992; Tornesello, Buonaguro, F.M., Beth-Giraldo, E., Kyalwazi, & Giraldo, 1992; Cupp, Malek, Goellner, Smith, & Espy, 1995; Gregoire, Cubilla, Reuter, Haas, & Lancaster, W.D. 1995; Cubilla, A.L., et. al., 1998, 2000; Picconi, et al., 2000; Bezerra, et al., 2001; Ferreux, et al., 2003; Gross, & Pfister, 2004; Senba, et al., 2006; Tornesello, et al., 2008). There are more than 40 different cancer-associated HPV types, but little is known about their geographic variation. HPV-16 is the most prevalent type in penile cancer (Rubin, et al., 2001; Haideman, et al., 2007; Pascual, et al., 2007; Tornesello, et al., 2008). However, in our investigation, HPV-18 is the most prevalent type in northern Thailand (Figure 1). Although it is unclear why the low-risk HPV-6 found in 54% of infections in northern Thailand has an important role in the oncogenic process of penile cancer, the synergistic action of HPV associated with multiple epidemiological cofactors such as life environment, personal hygiene, sexual habits, cultural practices, host genetic factors such as human leukocyte antigen (HLA) genes and others might activate oncogenes and/or mutant forms of oncosuppressor genes (Castellsague, Bosch, & Munoz, 2002; Hildesheim & Wang, 2002; Senba et al., 2006). HPV-6 and HPV-11 are detected in condyloma lesions (Figure 2).

Anal cancer is a rare malignancy of the anogenital tract. The anal cancer incidence in United State from 1996 to 2000 was 2.1 per 100,000 people, both male and female (Johnson, Madeleine, Newcomer, Schwartz, & daling, 2004). The primary cause of cervical cancer associated with HPV is the development of anal cancer. Anal intercourse is a risk factor for the development of both squamous cell carcinoma and cloacogenic anal cancer (Daling et al., 1987). HPV-16 was the most frequent HPV type detected, followed by HPV-18, regardless of gender. The high proportion of cancers with detectable HPV suggests that infection with HPV is a necessary cause of anal cancer, similar to the situation of cervical cancer (Daling, et al., 2004). A high prevalence of HPV infection in the anus has been reported in HIV infected men and women under immunosuppression. The incidence of anal cancer is much higher in HIV infected men who practice anoreceptive intercourse (Sierra, et al., 2006; Uronis, & Bendell, 2008). Relatively little is known about the epidemiology of anal infection among healthy women associated with cervical cancer. Factors for HPV infection in men are current and past sexual behavior, circumcision, status, lack of condom use, history of other sexually transmitted infections, cigarette smoking and education level (Daling, et. al., 2004; Nielson, et al., 2007).

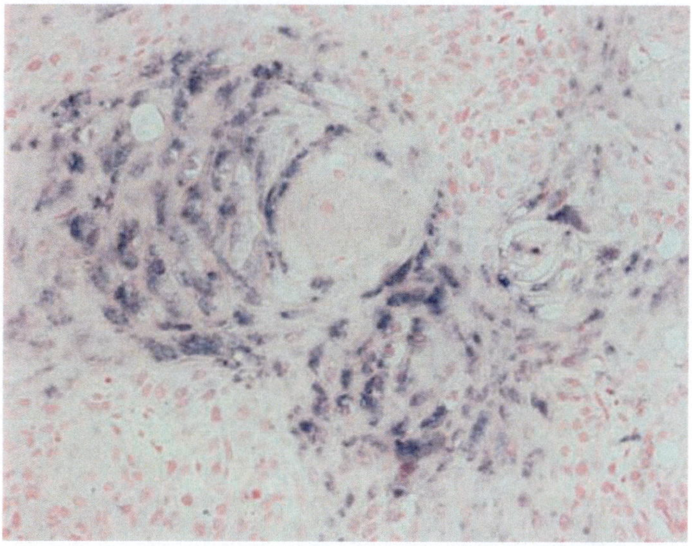

Figure 1. In situ hybridization for human papillomavirus signal exhibited on section of penile cancer.

Between 2000 and 2004, cancer of the oral cavity and pharynx was the 9th most common cancer among men in the United States, with an incidence rate of 16.0 per 100,000 (Espey, et al., 2007). Epidemiological data shows that the annual incidence of base of tongue and tonsil cancers in the USA increased by 2.1% and 3.9%, respectively, from 1973 to 2001 (Frisch, Hjalgrim, Jaeger A.B., & Biggar, R, 2000; Shiboski, Schmidt, & Jordan, 2005). This increase in the incidence of oropharyngeal squamous cell carcinoma (OSCC) was paralleled by an increase in certain sexual behaviors. Head and neck cancer is consistent with a role for genital HPV in the pathogenesis of OSCC. High-risk HPV-16 and HPV-18 infections have a strong association with oral cancer (Kreimer, Clifford, Boyle, & Francescgi, 2005; Syrjanen, 2005; Psyrri, & DiMaio, 2008). HPV-6 and HPV-11 has been detected in condilomatous oral lesions (Syrjanen, 2003). Sexual practices are typically associated with sexual transmission of the virus. Several studies suggest that oral HPV infection is sexually acquired. Oral-to-genital contact was strongly associated with OSCC, but mouth-to-mouth contacts cannot be ruled out in HPV transmission. Widespread oral sexual practices among adolescents may be contributing to HPV infection in the oral cavity, resulting in an increase of OSCC (D'Souza, et al., 2007). There is also an increased risk of HPV associated OSCC in individuals with a history of HPV associated with anogenital cancers and poor oral hygiene. HPV may be a significant contributing factor in oral cancer, but is not a necessity in all cases (Ha, & Califano, 2004). The most important risk factor for the development of oral cancer is tobacco and alcohol use (Scully, 2002; Warnakulasuriya, Sutherland, & Scully, 2005; Ogden, 2005; Tsantoulis, Kastrinakis, Tourvas, Laskaris, & Gorgoulis, 2007).

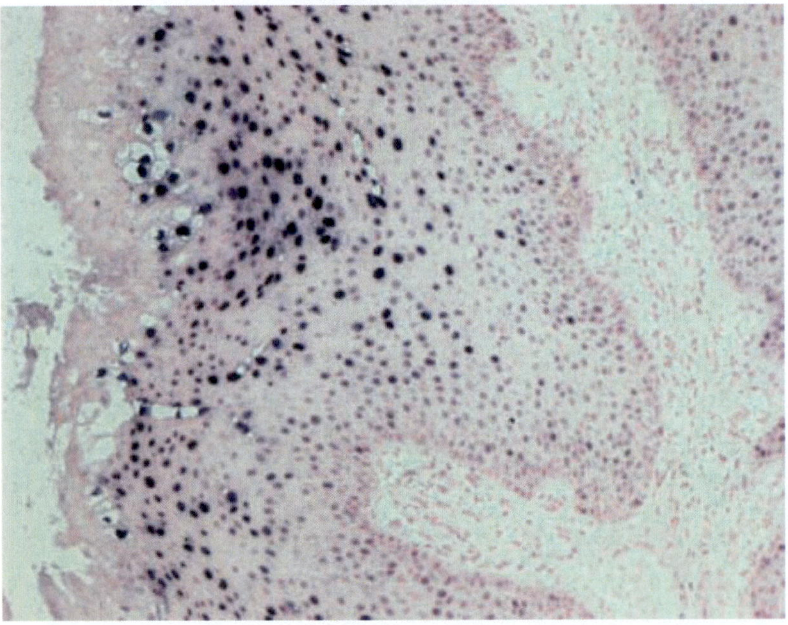

Figure 2. In situ hybridization for human papillomavirus signal exhibited on section of penile condyloma.

Nonmelanoma skin cancer (NMSC) is the most common cancer among Caucasians. The two main histological types of NMSC are squamous cell carcinoma (SCC) and basal cell carcinoma (BCC). The amount of HPV DNA seems to be higher in skin cancers of immunosuppressed patients than in those of the general population. In immunosuppressed patients, HPV prevalence is generally higher in SCC than in BCC (Pfister, 2003) There is increasing evidence reporting skin HPV types in the process of skin oncogenesis, but the role of HPV in the development of skin cancer is not clear. The most frequent manifestation of skin HPV infection is the development of benign self-limiting epithelia proliferations, namely warts. The relationship between HPV and NMSC is important. However, other cofactors are necessary for progression from skin HPV infection to cancer after more than a decade. A likely role for HPV infection in skin oncogenesis as a cofactor in association with ultraviolet (UV) has been proposed. Viruses transform keratinocytes by activation of cancer promoting genes and viral proteins may directly act as oncogenes. UV is one of the most important environmental oncogenesis factors for NMSC (Pfister, 2003; Akgul, Cooke, & Storey, 2006). HPV both stimulates cell proliferation and inhibits apoptosis, and additional genetic changes are induced by HPV infection and UV exposure resulting in cancer formation. Mutations induced in genomic and mitochondrial DNA by UV, makes it the most important etiological agent in the development of NMSC. HPV most likely contributes to NMSC development through a role as a co-factor. Skin cancer risk behaviors were infrequent use of sun-protective clothing and infrequent use of sunscreen (Coups, Manne, & Heckman, 2008).

HPV Genome and HPV Types

HPV are small DNA tumor viruses that induce hyperproliferative lesions in cutaneous and mucosal epithelia. The genome of HPV is a long circular double-stranded DNA molecule of approximately 8000 base pairs (bp) (Figure 3). The HPV genome is organized into three regions; the late gene (L1 and L2), the early gene (E1, E2, E3, E4, E5, E6, and E7), and the long control region (LCR). Viral genes are divided into early and late categories dependent on the time of expression (Munger, et al., 2004; Hebner, & Laimins, 2006; Zheng, & Baker, 2006; Narisawa-Saito, & Kiyono, 2007).

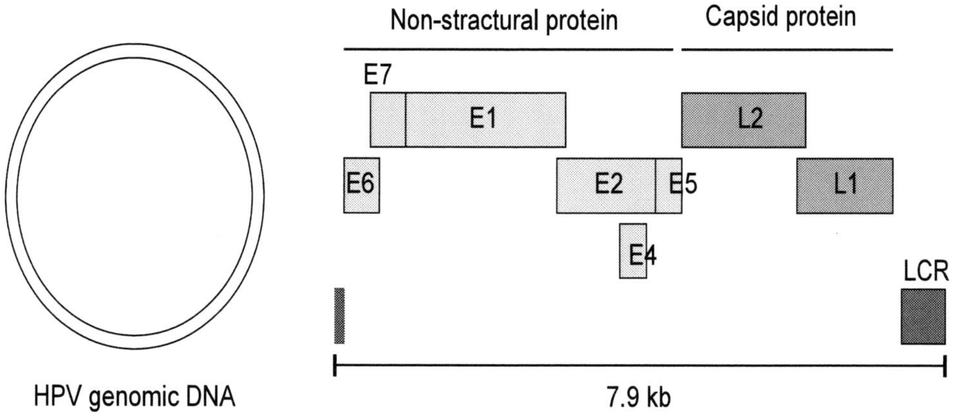

Figure 3. The HPV genome is a circular double-stranded DNA molecule of approximately 8000 base pairs. L1 is the major capsid protein and L2 is the minor capsid protein. The E1 to E7 encoding proteins are involved in viral replication and oncogenesis. The long control region (LCR) is an upstream regulatory region.

E1 and E2 encode proteins and completion of the viral life cycle due to initiation of viral replication and regulation of viral transcription. The E1 protein begins HPV DNA replication for binding to DNA polymerase. The E2 protein inhibits transcription of the early regions

(Ozbun, & Meyers, 1998). HPV associated with cervical cancer exhibits a loss of the expression of the viral E2 protein (Thierry, et al., 2004).

The E4 protein is expressed in the later stages of infection when complete virions are assembled. The E4 protein is considered to play an important role in the maturation and replication of the HPV viruses, and so is involved with the induction of the release of virions from infected cells (Doorbar, Ely, Sterling, Mclean, & Crawford, 1991; Brown, Fan, Jones, & Bryan, 1994).

The E5 protein has weak activity in maintaining the malignant transformation of the host cells, and interacts with various transmembrane proteins like receptors, such as epidermal growth factor, plated derived growth factor β, and colony stimulating factor-1 (Hwang, Nottoli, & Dimaio, 1995; Crusius, Auvinen, & Alonso, 1997).

Proteins E6 and E7 have been shown to induce immortalization and transformation of the host cells (Thomas, Pim, & Banks, 1999; Duensing, & Munger, 2004).

The late gene regions, L1 and L2 encode for viral capsid proteins during the late stages of virion assembly. The protein encoded by L1 is highly conserved among different HPV species. The minor capsid protein encoded by L2 has more sequence variation than that of the L1 protein (Florin, Sapp, Streeck, & Sapp, 2002; Modis, Trus, & Harrison, 2002; Becker, Florin, Sapp, Maul, & Sapp, 2004) (Table 2).

Table 2. Overview of HPV gene products

Gene product	Description
E1	Viral DNA replication and control of gene transcription
E2	Viral transcription and viral DNA replication
E3	
E4	Viral assembly
E5	Growth stimulation by interaction with growth factor receptors
E6	Cell immortalization due to p53-degradation, telomerase activation, anti-apoptosis effece, and induction of genomic instability
E7	Cell immortalization due to interaction with pRb, transactivation of E2F-dependent promoters, and induction of genomic instability
L1	Major capsid protein
L2	Minor capsid protein

E: early, L: late.

The early gene regions are encoding proteins, which are necessary for the replication of viral DNA and for the assembly of newly produced virus particles within infected cells. Early viral transcription is regulated by an element located in the non-encoding region proximal to the open reading frames (ORF) called the long control region (LCR) or the upstream regulatory region (URR). The LCR contains cis-elements required for regulation of gene expression, replication of the genome, and its packaging into virus particles. The LCR has numerous binding sites due to many repressor and activators of transcription. It has been suggested that the LCR plays a part in determining the characteristics of HPV types (Morris,

Dent, Ring, & Latchman, 1993a; Morris, Ring, Lillycrop, & Latchman, 1993b; Tan, Leong, Walker, & Bernard, 1994; Kanaya, Kyo, & Laimins, 1997).

More than 100 HPV genotypes have been reported and approximately 40 of them infect the genital mucosa (Bernard, 2005). HPV are classified into low-risk and high-risk groups based on the potential for induction of malignant formation. Low-risk HPV types typically induce benign tumors and are rarely associated with malignant tumors. In contrast, high-risk HPV types induce malignancy. Thus, HPV are classified into low-risk types (6, 11, 26, 30, 34, 40, 42, 43, 44, 53, 54, 55, 57, 61, 62, 64, 67, 70, 71, 72, 73, 74, 79, 81, 82, 83, 84) or high-risk types (16, 18, 31, 33, 35, 39, 45, 51, 52, 54, 56, 58, 59, 66, 68, 69). The high-risk types are associated with more than 90% of cervical cancers (Table 3). Moreover, three main groups of HPV are classified depending on their localization, as follows; skin HPV types (1, 5, 8, 14, 20, 21, 25, 47), skin and mucosal types (2, 57), and mucosal types (6, 11, 13, 16, 31, 33, 35, 39, 42, 44, 45, 49, 51, 52, 56, 58, 68) (Gross, & Pfister, 2004; Villiers, Fauquet, Broker, Bernard, & Hausen, 2004; Cogliano, Baan, Straif, Grosse, Secretan, & Ghissassi, 2005) (Table 4). Among the high-risk strains, HPV-16 and HPV-18 are the most closely associated with cervical and penile cancer. HPV-6 and HPV-11 are typically found in genital warts and condyloma acuminata of the genital organs (Bernard, 2005).

Table 3. Classification of low- and high-risk HPV types

	HPV types
high-risk	16, 18, 31, 33, 35, 39, 45, 51, 52, 54, 56, 58, 59, 66, 68, 69
low-risk	6, 11, 26, 30, 34, 40, 42, 43, 44, 53, 54, 55, 57, 61, 62, 64, 67, 70, 71, 72, 73, 74, 79, 81, 82, 83, 84

Table 4. Localization of HPV types

Skin types	1, 5, 8, 14, 20, 21, 25, 47,
Skin and mucosal types	2, 57,
Mucosal types	6, 11, 13, 16, 31, 33, 35, 39, 42, 44, 45, 49, 51, 52, 56. 58, 68,

HPV Life Cycle

The HPV life cycle is tightly adapted to their host tissue, the differentiating epithelial cells of skin or mucosa. Normal squamous epithelial cells grow as stratified epithelium. Amplifying stem cells in the basal layer differentiate, and one of the daughter cells migrates upward and begins to undergo terminal differentiation. The other daughter cells remain in the basal layer as a slow-cycling and self-renewing population. HPV virions initially infect the basal layers of the epithelia, where they bind to and enter into cells through small wounds. The mechanism by which they actively invade cells involves the interaction with an as yet unidentified receptor. HPV virions migrate to the nucleus as episomes and early HPV promoter is activated.

The replication cycle within the epithelium can be divided into the latent and the lytic infections. In the latent infection, the viral genome is replicated to low copy numbers, approximately 10 to 200 genomes per cell within the initially infected the basal cells (Doorbar, 2005). The early HPV genes E1 and E2 are essential for DNA replication in the basal cells and for its segregation. The infected basal cells can be maintained in the lesion for a long period. At the viral persistence state, the immune system may confine the infection to these basal cells. In the lytic infection, when infected daughter cells migrate to the upper layers of the epithelium, viral late gene products are produced to initiate the vegetative phase of the HPV life cycle, resulting in high-level amplification of the viral genome. The HPV virions are synthesized at the upper layers of stratified squamous epithelia. For multiplication of viral DNA, the early genes E5, E6 and E7 are thought to coordinate a suitable environment in the host cells. The late promoter activates the late gene products, resulting in the vegetative phase of the HPV life cycle leading to high level amplification of the viral genome. The HPV virions replicate in the superficial layers of the epithelium, and consequently the progeny HPV virions are released from the cell to re-initiate infection (Fehrmann, & Laimins, 2003; Doorbar, 2005) (Figure 4).

By epidemiological and experimental studies, long persistent infection of the high-risk HPV is not enough to induce cancers, however, additional genetic activation and/or suppression leads to the formation of cancers. Increased levels of TERT, PIK3CA, C-MYC, ErbB2, MYOD1, CCND1, and ras, and decreased expression of PTEN, and TSLC1 have been reported in cervical cancers (Zhang, et al., 2002; Narayan, et al., 2003; Widschwendter, et al., 2004; Steenbergen, et al., 2004). ErbB2 protein expression level is also regulated by p53 degradation and interference with this by E6/E6AP complexes contributes to cervical oncogenesis.

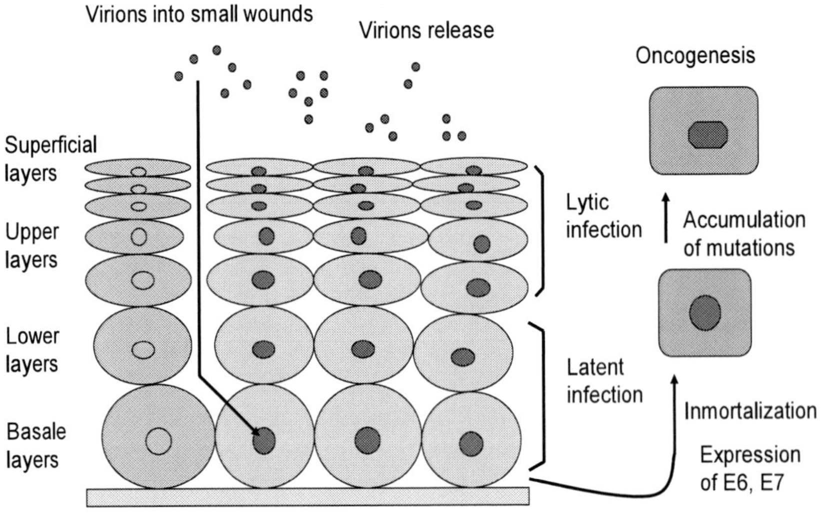

Figure 4. Latent infection: HPV initially infects the basal layer cells. HPV migrate to the nucleus and the early HPV promoter is activated. Few viral DNA synthesize in the infected cells. Lytic infection: Many viral DNA synthesize in the upper layer cells, and release from the superficial layer cells.

Elemental Function of E7 and E6 Oncogenesis

HPV onco-proteins E6 and E7 are essential factors for HPV oncogenesis. E7 and E6 react with the tumor suppressor gene p53 and pRb in host cell protein, resulting in induced cellular immortalization, transformation and carcinogenesis, due to their interference with cell cycle and apoptosis control (Munger, et al., 2004). E7 and E6 induced genetic instability leads to the activation of oncogenes and inactivation of tumor suppressor genes.

The HPV viral oncogenes, E7 and E6, have been shown to be the main contributors to the development of HPV induced cancers, probably due to integration of the viral genes in the host cell genome. Inactivation of tumor suppressor p53 and pRb is a common event in the carcinogenesis of human cells. In HPV infection, the significant interactions are with pRb and p53 proteins, which are important molecules in the cell cycle and apoptosis control. Remarkably, pRb and p53 proteins are mutated in many human cancers. Both E7 and E6 HPV oncogenes interact with pRb and p53, which inhibit the activities of these tumor suppressors. The cell cycle in S phase would normally lead to apoptosis by the action of pRb. However, in HPV infected cells, this process is counteracted by the viral E6 protein, which targets p53 for proteolytic degradation. High-risk HPV cause persistent infections which induce the initiation and progression of malignant tumor formation. As an aberration of virus infection, the constant activity of the viral proteins E7 and E6 leads to increasing genomic instability, and accumulation of oncogene mutations, resulting in cancer formation. Specifically, those involving tumor suppressor genes can give rise to an immortalization associated with activation of telomerase. Extensive studies have linked the efficiency of mucosal HPV types in promoting cancer development to the viral E6 and E7 proteins. The E6 and E7 genes of HPV are thought to play causative roles, since E6 promotes the degradation of p53 through its interaction with E6AP, an E3 ubiquitin ligase, whereas E7 binds to pRb and disrupts its complex formation with E2F transcription factors (Huibregtse, Scheffner, & Howley, 1993; Scheffner, Huibregtse, Vierstra, & Howley, 1993). In addition, E6 and E7 cause degradation of the cellular genes controlling the G2/M phase transition and progression (Walbooners, et al., 1999; Patel, D., Incassati, A., Wang, N., & McCance, D.J., 2004).

E6 and E7 proteins may immortalize various types of human cells independently, however, their cooperative interaction leads to substantially enhanced immortalization efficiency. E7 binds to cyclin-kinase complexes and cyclin-dependent kinase inhibitors. E7 directly binds cyclin A/CDK2 as well as indirectly interacting with the cyclin E/CDK2 complex via the p107 protein of the Rb family, which acts to increase cyclins A and E in the epithelial cells. E7 also binds the cyclin kinase inhibitors p27 and p21 (Tommasino, et al., 1993; McIntyre, Ruesch, & Laimins, 1996; Zerfass-Thome, et al., 1996; Funk, et al., 1997; Ruesch, & Laimins, 1998). The functional inactivation of the cyclin-dependent kinase inhibitors p21 and p27 by E7 are probably the key events in this cooperation (Hausen, 2000) (Figure 5).

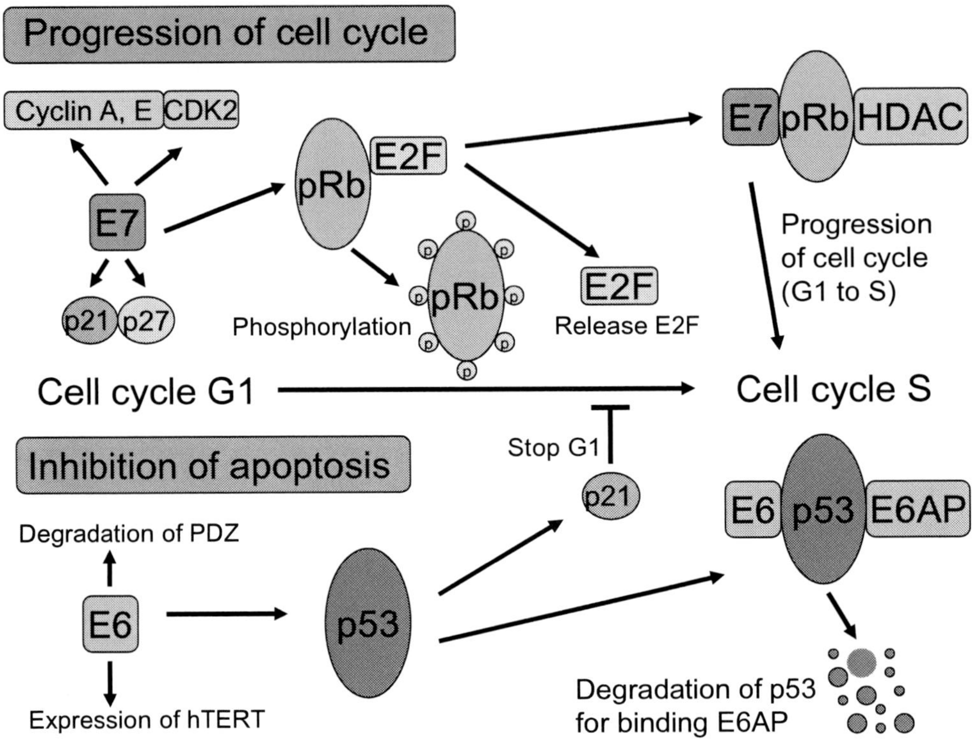

Figure 5. Essential mechanisms for the interaction of E7 with pRb and interaction of E6 with p53.

FUCTION OF E7 ONCOGENESIS

E7 is an oncogene that mediates the initiation of DNA synthesis and transforms cells into continuously growing cells. It thereby induces genetic abnormalities that enhance the likelihood of malignant progression. The E7 protein can interact with pRb, an important negative regulator of entry into the S phase of the cell division cycle. It is thought that the high-risk HPVs have a unique ability to induce the proteolytic degradation of pRb. In the hypophosphorylated state, combined E7 and pRb activates the E2F transcription factor, which triggers the expression of proteins necessary for DNA replication and cell cycle progression. Phosphorylation of pRb by G1 cyclin-dependent kinases releases E2F leading to cell cycle progression into the S phase. Because E7 is able to bind to unphosphorylated pRb, it may prematurely induce cells to enter the S phase by disrupting pRb-E2F complexes (Boyer, Wazer, & Band, 1996; Jones, Thompson, & Munger, 1997; Berezutskaya, Yu, Morozov, Raychaudhuri, & Bagchi, 1997; Nguyen, Westbrook, & McCance, 2002).

The $p16^{INK4a}$ is a cyclin-dependent kinase inhibitor which prevents the phosphorylation of pRb family members. Normally, overexpression of $p16^{INK4a}$ results in inhibition of E2F dependent transcription and inhibition of cell cycle progression at the G1 to S checkpoint (Zhang, Postigo, & Dean, D.C., 1999). The repression of $p16^{INK4a}$ gene expression by hypermethylation or mutation is a common occurrence in cancer. There is a close association between p16 overexpression and high-risk HPV infection. Therefore, overexpression of $p16^{INK4a}$ is suggested to be a useful marker for evaluating HPV activity in cancer lesions (Sano, Oyama, Kashiwabara, Fukuda, & Nakajima, 1998; Ferreux, et al., 2003).

E7 protein function enables HPV replication in the upper layers of the epithelium. The E7 oncoprotein contains approximately 98 amino acids and is a small nuclear phosphoprotein separated into three conserved regions denoted in an analogous fashion to adenovirus E1A as CR1, CR2 and CR3 (Barbosa, et al., 1990; Dyson, Guida, Munger, & Harlow, 1992). E7 is known to bind to pRb, and its family members, p107 and p130, via a LXCXE (where X represents any amino acid) binding motif conserved in its CR2 region. The LXCXE motif necessary for pRb binding is in the CR2 region of E7 (Chellappan, et al., 1992; Gonzalez, Stremlau, He, Basile, & Munger, 2001; Munger et al., 2001).

In addition to the inactivation of pRb family members, numerous functions of E7 have been reported. Histone deacetylases (HDAC) and transcriptional co-repressors have been reported to associate with E7 via Mi2 to promote cell growth. The interaction of E7 with cyclin-dependent kinase (CDK)2/cyclin A, as well as CDK2/cyclin E, has also been reported. These cyclin-kinase complexes exhibit kinase activities that can phosphorylate the pRb proteins. Furthermore, E7 binds the CDK-inhibitor (CKI), p27, and CKIp21, confirming the abrogation of cell-cycle inhibition. For evasion of the immune system, E7 also interacts with p48, as well as IRF1. Recently, p600 has been identified as a cellular target of E7 that contributes to anchorage-independent growth and cellular transformation. Because knockdown of p600 sensitizes cells to apoptosis induced by cell detachment irrespective of the presence of E7, it has been suggested as a novel target of cancer treatment. Binding of E7 with both the 35-kDa catalytic and 65-kDa structural subunits of PP2A has been reported, resulting in the sequestration of these subunits and inhibition of their interaction with PKB/Akt, thereby maintaining its signaling by blocking its dephosphorylation. Despite these multifunctional properties of the E7 protein, however, the expression of a mutant form of pRb (DeltaLXCXE) that is selectively defective for binding E7 revealed that most the effects of E7 on epidermal differentiation are indeed due to pRb inactivation (Hausen, 2000; 2002).

Function of E6 Oncogenesis

The most manifest function of the E6 protein is to promote the degradation of p53 under E6 protein associated with E6AP protein and an E3 ubiquitin ligase. The affinity of E6AP for p53 is likely to be modified by the association with E6. p53 is a tumor suppressor gene involved in apoptosis after DNA damage, and regulation of both the G1/S and G2/M cell cycles. HPV-infected cells induce inhibition of apoptosis by E6 inactivation of p53. In addition, E6 interferes with other pro-apoptotic proteins, such as Bak, FADD and procaspase 8 (Thomas, Pim, & Banks, 1999; Garnett, Filippova, & Duerksen-Hughes,, 2006).

The product of the Notch1 gene has been identified as a novel target of p53. Notch1 function has been observed as an oncogene in the development of human T-cell leukemia, but also acts as a determinant of keratinocyte differentiation and tumor suppressor in the epidermis. Notch1 expression leads to squamous cell carcinomas. Induction of Notch1 through p53 occurs in response to genotoxic stress. Therefore, its down regulation through p53 with E6/E6AP has been revealed as a novel tumor suppressor mechanism that blocks the development of HPV induced oncogenesis which leads to HPV related cancer (Lefort, et al., 2007; Yugawa, et al., 2007).

An important function of the E6 protein is the activation of telomerase in infected cells. Target proteins of E6 might contribute to cellular transformation with telomerase. Human telomerase is a ribonucleoprotein complex composed of at least the reverse catalytic transcriptase (hTERT) and RNA component (hTR). The hTERT is expressed in specific germ-line cells, proliferative stem cells of renewal tissues, and cancer cells. In normal cells, the expression of hTERT reconstitutes telomerase activity and suppresses senescence. High telomerase activity is observed in the most human cancer cells, which strongly suggests an important role in oncogenesis. Telomere shortening due to incomplete replication by DNA polymerase during mitosis is an important factor that limits the lifespan of normal cells. However, E6 induces telomerase activity, resulting in a contribution to the immortalization of epithelial cells by maintaining telomere length. The E6 protein has been shown to increase telomere length in epithelial cells by activating the catalytic subunit of the enzyme telomerase, hTERT. E6 and Myc interaction has been observed to activate the telomerase reverse transcriptase promoter (Klingelhutz, Foster, & McDougall, 1996; Oh, Kyo, & Laimins, 2001; McMurray, & McCance, 2003; Veldman, Liu, Yuan, & Schlegel, 2003; Gewin, Myers, Kiyono, & Galloway, 2004). NFX1-91 is an identified novel cellular repressor of the hTERT promoter. The hTERT promoter associated with Myc leads to increased hTERT expression (Katzenellenbogen, et al., 2007).

High-risk HPV E6 has been shown to interact with PSD95/Dlg/Zo-1 (PDZ) domain-containing proteins leading to their degradation. The PDZ domain itself is composed of 80 to 100 amino acid residues and mediates binding to a C-terminal motif or other PDZ domains within its interaction partners. The majority of these binding partners are transmembrane proteins located at the plasma membrane including transmembrane receptors, and channel proteins. (Jelen, Oleksy, Smietana, & Otlewski, 2003). E6 targets several other cellular proteins for degradation and possesses the ability to bind and degrade p53. This is important for cell transformation because PDZ-domain-containing proteins are involved in a variety of cellular functions such as cell signaling, cell adhesion, ion exchange, and tight-junction integrity. PDZ proteins have also been found to be target proteins by oncoproteins from other tumor viruses, such as human T cell lymphoma virus Tax and adenovirus E4 orf1. E6 interacts with several PDZ proteins, including MAGUKs (Dlg, Scribble, MAGI-1, MUPP1) and TIP-1, via a C-terminal PDZ binding motif on the E6 protein. E6 binding to these proteins appears to be particularly important for transformation and oncogenesis in cultured cells and transgenic mice. Recently, several PDZ-domain-containing proteins have been identified as targets of E6 proteins. E6-induced degradation of these proteins potentially causes loss of cell-cell contacts mediated by tight junctions. E6 also contributes to the loss of cell polarity in HPV-associated cancers. The tumor suppressor properties of some of these proteins against the development of HPV associated cancer have been reported. The targeting of PDZ domain containing proteins may lead to oncogenic transformation associated with molecular changes (Glaunsinger, Lee, Thomas, Banks, & Javier, 2000; Nguyen, Nguyen, Lee, Griep, & Lambert, 2003; Watson, Thomas, Banks, & Roberts, 2003; Lee, & Laimins,

2004; Favre-Bonvin, Reynaud, Kretz-Remy, & Jalinot, 2005; Storrs, & Silverstein, 2007) (Figure 6).

The tumor suppressor gene TSC2 product, Tuberin, has been proposed as a possible target of E6, implying a contribution to E6-induced oncogenesis. Because TSC complexes play an important regulatory role in the survival signaling involving mTOR, further exploration of this possibility is needed. More recently, induction by E6 of E2F-responsive genes, MCM7 and cyclin E, has been reported, implying the existence of dysregulation of the p16/pRb pathway with mechanisms distinct from those involving E7 (Shai, Brake, Somoza, & Lambert, 2007).

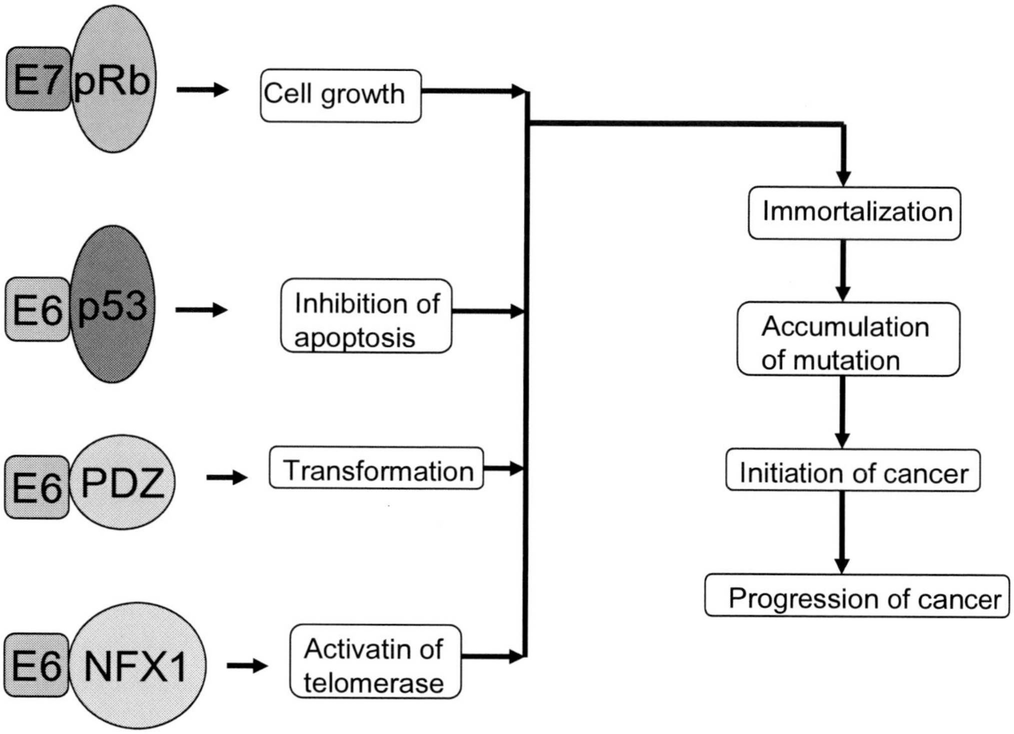

Figure 6. E7 and E6 oncoproteins of the high-risk HPV can inactivate tumor suppressors pRb and p53, respectively. NFX is a recently identified novel cellular repressor of the hTERT promoter. PDZ-domain-containing proteins lead to degradation of p53, which is important for cell transformation.

NF-κB PROMOTION IN INFLAMMATION AND CANCER DEVELOPMENT

The causal relationship between chronic inflammation and cancer is widely accepted. Numerous investigations have identified nuclear factor-κB (NF-κB) as an important modulator in driving chronic inflammation to cancer. This transcription factor is indispensable for the malignant progression of transformed cells associated with various inflammatory cells and a network of signaling molecules. The expression and the function of numerous cytokines, chemokines, growth factors, and survival factors are NF-κB-dependent. NF-κB activation has been implicated in a variety of processes related to transformation and oncogenesis, including proliferation, migration, angiogenesis, and prevention of apoptosis. NF-κB activity is involved in the regulation of the angiogenesis process for cancer cell growth and invasiveness. VEGF is the main member of angiogenic factors, and is under the transcriptional control of NF-κB (Kiriakidis, et al., 2003). Increased NF-κB activity is associated with many cancers, including cancers associated with viral infections. NF-κB dependent proliferation and protection from apoptosis are likely to have significant effects on the oncogenesis of HPV associated with cancers. HPV E6 and E7 positive cells have shown IL-1β induced NF-κB activation and exhibit elevated levels of NF-κB components (Havard, Delvenne, Frare, Boniver, & Giannini, 2002). E6 rather than E7 expression was found to be associated with the nuclear location of these components.

NF-κB activity is modulated for many different viruses, such as HIV-1, HTLV-1, EBV, HBV, adenovirus, and HPV (Nees, et al., 2001; Spitkovsky, Hehner, Hofmann, Moller, & Schmitz, 2002; Havard, Delvenne, Frare, Boniver, & Giannini, 2002; Havard, Rahmouni, Boniver, J, & Delvenne, 2005; Mishra, Varghese, Saluja, & Das; 2006; James, Lee, Aloysius, & Klingelhutz, 2006). It is frequently reported that HPV encoded E6 and E7 oncoproteins are important regulatory proteins in host cells, which are associated with the transcriptional activity of NF-κB. A fraction of the E7 protein is found in association with the IκB kinase complex and attenuates induced kinase activity of IκB kinase α (IKKα) and IKKβ, thus resulting in impaired IκBα phosphorylation and degradation. While E7 obviates IKK activation in the cytoplasm, the E6 protein reduces NF-κB p65-dependent transcriptional activity within the nucleus. It is suggested that HPV oncogene-mediated suppression of NF-κB activity contributes to HPV escape from the immune system (Spitkovsky, Hehner, Hofmann, Moller & Schmitz, 2002).

PDZ domain-containing proteins have been identified as binding partners for the oncoprotein E6 of the high-risk type HPV. The PDZ binding motif of E6 is required for activation of NF-κB and the non-PDZ domain containing protein by E6 may be necessary for NF-κB activation (James, Lee, & Klingelhutz, 2006). Tumor necrosis factor (TNF), other cytokines, and growth factors stimulate the activation of NF-κB and the transcription of anti-apoptotic target genes, including the inhibitor of apoptosis proteins XIAP, cIAP-1, and cIAP-2 (Gaur, & Aggarwal, 2003). Prevention of DNA damage or cytokine induced apoptosis by activation of NF-κB may be a major obstacle in the treatment of HPV associated cancers. NF-κB activation in HPV infected cells likely plays a role in the proliferative capacity of the cells in addition to protection from apoptosis, which is reflected in the diminished ability of E6 mutants that are not able to bind PDZ domains or activate NF-κB to promote oncogenesis. Investigation of the potential role of NF-κB activation in the pathogenesis of HPV associated cancers is of significant importance with regards to the treatment and prevention for viral cancers (James, Lee, & Klingelhutz, 2006).

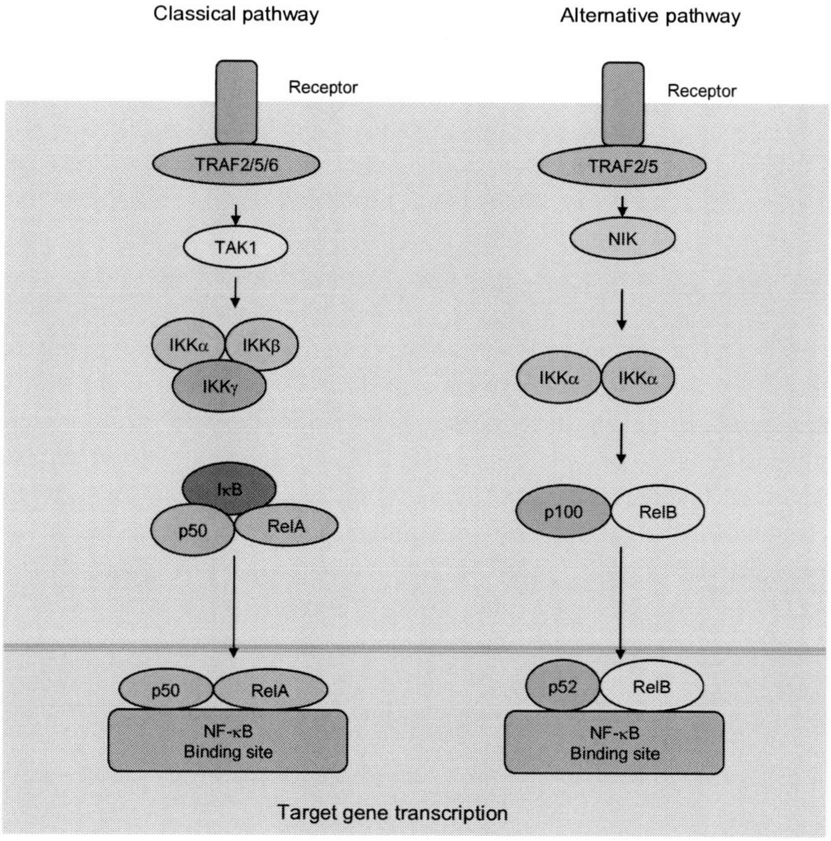

Figure 7. Two signalling pathways lead to the activation of NF-κB. The classical pathway is triggered by proinflammatory cytokines, bacterial-wall components, viruses and genotoxic stress. The alternative pathway is triggered by TNF-family members.

The IκB family regulates the activity of NF-κB complexes in the cytoplasm. NF-κB translocates to the nucleus to interact with the κB sites located in the regulatory regions of target genes. The NF-κB/Rel superfamily of eukaryotes is a group of ubiquitously expressed, pleiotropic, and transcriptional regulators. The Rel family is classified into 2 categories. One group contains RelA (p65), RelB and c-Rel, which do not require proteolytic processing to produce mature forms. The other group contains NF-κB1 (p105) and NF-κB2 (p100), which do require processing to produce the mature p50 and p52 proteins, respectively. The most abundant NF-κB dimmer is p50/RelA, which is bound to the inhibitory protein IκB. Nuclear translocation of NF-κB is separated into two NF-κB activation pathways, such as the classical pathway and the alternative pathway. The classical pathway is triggered by bacterial and viral infections. The IKK complex is activated by pro-inflammatory cytokines, and is composed of two catalytic subunits, IKK-α and IKK-β, and a regulatory subunit, IKK-γ. The IKK complex phosphorylates NF-κB bound IκBs, thus targeting them for proteasomal degradation and liberating NF-κB dimmers that are composed of RelA and p50 of the NF-κB mediated gene induction response. The p50-RelA dimmers lead to increased transcription of genes. The alternative NF-κB activation pathway involves the upstream kinase NF-κB inducing kinase (NIK) that activates dependent on IKK-α, independently of either IKK-β or IKK-γ leading to the phosphorylation and specific serine residues in the C-terminal half of p100, in response to certain members of the TNF family. This phosphorylation leads to polyubiquitination dependent degradation of the C-terminal half of p100 to generate p52-REL-B heterodimmers, which activate the target gene. The two pathways for switch have different gene sets, resulting in different immune function (Vancurova, Wu, Miskolici, & Sun, 2002; Karin, 2005; 2006; Basseres, & Baldwin, 2006; Inoue, Gohda, Akiyama, & Semba, 2007) (Figure 7).

The relationship between chronic inflammation and cancer progression has long been known. NF-κB activation and subsequent production of cytokines, chemokines, growth factors, and inhibition of apoptosis proteins have been observed in cancer progression. Toll-like receptors (TLRs) are possible signal initiators for NF-κB activation due to inflammation induced carcinogenesis. In most cancer cells, the activation of NF-κB plays an important role in cancer initiation, progression, and metastasis. TLRs are among the major activators of NF-κB and are the front line receptors in response to microbial infection. The activation of NF-κB through stimulated TLRs in local chronic inflammation may serve as an initiator. NF-κB is constitutively activated during human cervical cancer progression (Chen, Alero, Silasi, & Mor, 2007; El-Omar, Ng, & Hold, 2008; Chen, Alvero, Silasi, Steffensen, & Mor, 2008) (Figure 8).

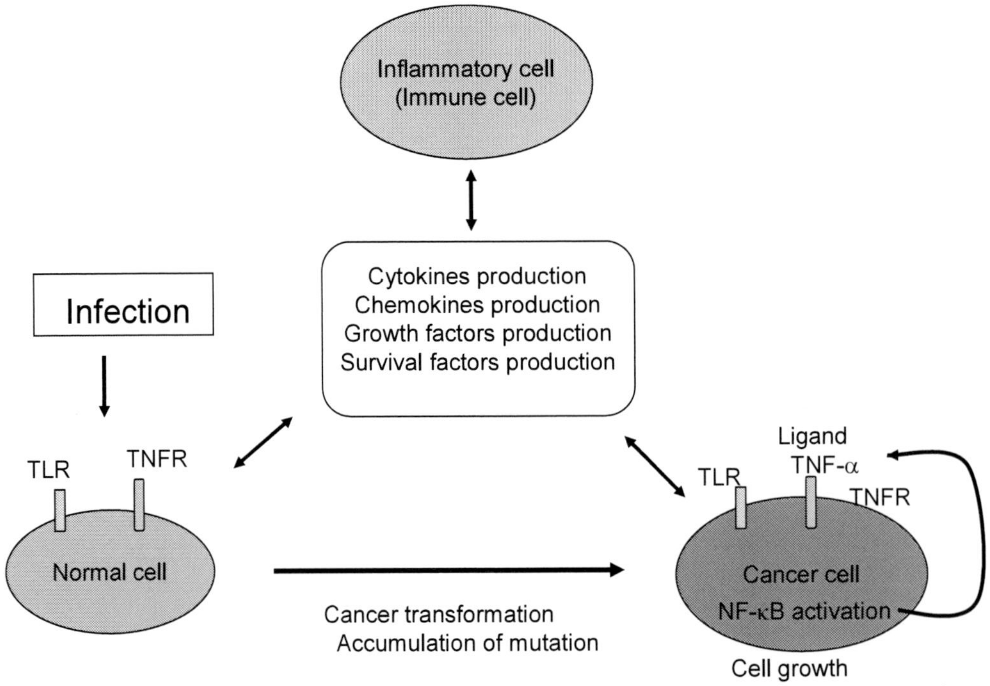

Figure 8. Relationship between chronic inflammation and cancer demonstrated link among Toll-like receptor (TLR), inflammation, and oncogenesis. TLR activates NF-κB, resulting in the production of inflammatory cytokines, chemokines, growth factors, survival factors, which could enhance malignant transformation.

Risk Factors and Cofactors for HPV Infection

Previous studies have associated environmental factors and specific aspects of lifestyle with HPV associated cancer. HPV infections initiate and progress to genital cancers after more than a decade, but oncogenic cofactors are necessary for the development of HPV related cancer. The ability of the high-risk HPV types to transform and immortalize infected cells has been established. Although HPV infection is strongly correlated with genital cancers, it is not a sufficient cause in itself. Multiple steps as well as multiple factors may be necessary in the development of cervical cancer. It has long been understood that cancer is a multi etiological disease, and the time lag between HPV infection and the diagnosis of cancer also implies the action of co-factors. Long-term use of hormonal contraceptives, high parity, tobacco smoking, and co-infection with HIV have been identified as established cofactors; co-infection with *Chlamydia trachomatis* (CT) and herpes simplex virus type-2 (HSV-2), immunosuppression, and certain dietary deficiencies are other probable cofactors. Genetic and immunological host factors and viral factors other than type, such as variants of type, viral load and viral integration, are likely to be important but have not been clearly identified (Khare, Pater, Tang, & Pater, 1997; Kjellberg, et al., 2000; Koskela et al., 2000; Hidesheim, & Wang, 2002; Schiffman, & Castle, 2003; Haverkos, 2004; Madkan, et. al., 2007). An oncogenetic role for HPV infection in the development of NMSC has been proposed; along with cofactors such as solar ultraviolet (UV) irradiation as the risk factor for NMSC development, as HPV-associated NMSC develops predominantly at sites exposed to the sun (Pfister, 2003; Akgul, Cooke, & Storey, 2006).

References

Aggarwal, B.B., Shishodia, S., Sandur, S.K., Pandey, M.K., & Sethi, G. (2006). Inflammation and cancer: How hot is the link? *Biochem. Pharmancol., 72*, 1605-1621.

Akgul, B., Cooke, J.C., & Storey, A. (2006). HPV-associated skin disease. *J. Pathol., 208*, 165-175.

Balkwill, F., & Mantovani, A. (2001). Inflammation and cancer: Back to Virchow? *Lancet, 357*, 539-545.

Barbosa, M.S., Edmonds, C., Fisher, C., Schiller, J.T., Lowy, D.R., & Vousden, K.H. (1990). The region of the HPV E7 oncoprotein homologous to adenovirus E1a and Sv40 large T antigen contains separate domains for Rb binding and casein kinase II phosphorylation. *EMBO J., 9*, 153-160.

Basseres, D.S., & Baldwin, A.S. (2006). Nuclear factor-κB and of κB kinase pathways in oncogenic initiation and progression. *Oncogene, 25*, 6817-6830.

Becker, K.A., Florin, L., Sapp, C., Maul, G.G., & Sapp, M. (2004). Nuclear localization but not PML protein is required for incorporation of the papillomavirus minor capsid protein L2 into virus-like particles. J. *Virol., 78*, 1121-1128.

Bekkers, R.L., Massuger, L.F., Bulten, J., & Melchers, W.J. (2004). Epidemiological and clinical aspects of human papillomavirus detection in the prevention of cervical cancer. *Rev. Med. Virol., 14*, 95-105.

Bernard, H.U. (2005). The clinical importance of the nomenclature, evolution and taxonomy of human papillomaviruses. *J. Clin. Virol., 32* Suppl 1, S1-S6.

Bezerra, A.L., Lopes, A., Satiago, G.H., Ribeiro, K.C., Latorre, M.R., Villa, L.L. (2001). Human papillomavirus as a prognostic factor in carcinoma of the penis: Analysis of 82 patients treated with amputation and bilateral lymphadenectomy. *Cancer, 91*, 2315-2321.

Berezutskaya, E., Yu, B., Morozov, A., Raychaudhuri, P., & Bagchi, S. (1997). Differential regulation of the pocket domains of the retinoblastoma family proteins by the HPV16 E7 oncoprotein. *Cell Growth Differ., 8*, 1277-1286.

Bosch, F.X., Manos, M.M., Munoz, N., Sherman, M., Jansen, A.M., Peto, J., Schiffman, M.H., Moreno, V., Kurman, R., & Shah, K.V., (1995). Prevalence of human papillomavirus in cervical cancer: A worldwide perspective. *J. Natl. Cancer Inst., 87*, 796-802.

Boyer, S.N., Wazer, D.E., & Band, V. (1996). E7 protein of human papilloma virus-16 induces degradation of retinoblastoma protein through the ubiquitin-protesome pathway. *Cancer Res., 56*, 4620-4624.

Brown, D.R., Fan, L., Jones, J., & Bryan, J. (1994). Colocalization of human papillomavirus type 11 E1E4 and L1 proteins in human foreskin implants growth in athymic mice. *Virology, 201*, 46-54.

Bryan, J.T., Taddeo, F., Skulsky, D., Jansen, K.U., Frain, B.M., Qadadri, B., & Brown, D.R., (2006). Detection of specific human papillomavirus types in paraffin-embedded sections of cervical carcinomas. *J. Med. Virol., 78*, 117-124.

Castellsague, X., Bosch, F.X., & Munoz, N. (2002). Environmental co-factors in HPV carcinogenesis. *Virus Res., 89*, 191-199.

Chellappan, S., Kraus, V.B., Kroger, B., Munger, K., Howley, P.M., Phelps, W.C., Nevins, J.R. (1992). Adenovirus E1A, simian virus 40 tumor antigen, and human papillomavirus E7 protein share the capacity to disrupt the interaction between transcription factor E2F and the retinoblastoma gene product. *Proc. Natl. Acad. Sci. USA, 89*, 4549-4553.

Chen, R., Alvero, A.B., Silasi, D.A, & Mor, G. (2007). Inflammation, cancer and chemoresistance: Taking advantage of the Toll-like receptor signaling pathway. *Am. J. Reprod. Immunol., 57*, 93-107.

Chen, R., Alvero, A.B., Silasi, D.A., Steffensen, K.D., & Mor, G. (2008). Cancers take their Toll-the function and regulation of Toll-like receptors in cancer cells. *Oncogene, 27*, 225-233.

Cogliano, V., Baan, R., Straif, K., Grosse, Y., Secretan, B., El Ghissassi, F. (2005). Carcinogenicity of human papillomaviruses. *Lancet Oncol., 6*, 204.

Coups, E.J., Manne, S.L., & Heckman, C.J. (2008). Multiple skin cancer risk behaviors in the U.S. population. *Am. J. Prev. Med., 34*, 87-93.

Coussens, L.M., & Werb, Z. (2002). Inflammation and cancer. *Nature, 420*, 860-867.

Crusius, K., Auvinen, E., & Alonso, A. (1997). Enhancement of EGF-and PMA-mediated MAP kinase activation in cells expressing the human papillomavirus type 16 E5 protein. *Oncogene, 15*, 1437-1444.

Cubilla, A.L., Reuter, V.E., Gregoire, L., Ayala, G., Ocampos, S., Lancaster, W.D., Fair, W. (1998). Basaloid squamous cell carcinoma: A distinctive human papilloma virus-related penile neoplasm. A report of 20 cases. *Am. J. Surg. Pathol., 22*, 755-761.

Cubilla, A.L., Velazques, E.F., Reuter, V.E., Oliva, E., Mihm, M.C., Jr., Young, R.H. (2000). Warty (condylomatous) squamous cell carcinoma of the penis. *Am. J. Surg. Pathol., 24*, 505-512.

Cupp, M.R., Malek, R.S., Goellner, J.R., Smith, T.F., Espy, M.J. (1995). The detection of human papillomavirus deoxyribonucleic acid in intraepithelial, in situ, verrucous and invasive carcinoma of the penis. *J. Urol., 154*, 1024-1029.

Daling, J.R., Weiss, N.S., Hislop, T.G., Maden, C., Coates, R.J., Sherman, K.J., Ashley, R.L., Beagrie, M., Ryan, J.A., & Corey, L. (1987). Sexual practices, sexually transmitted diseases, and the incidence of anal cancer. *N. Engl. J. Med., 317*, 973-977.

Daling, J.R., Madeleine, M.M., Johnson, L.G., Schwartz, S.M., Shera, K.A., Wurscher, M.A., Carter, J.J., Porter, P.L., Galloway, D.A., McDoogall, J.K. (2004). Human papillomavirus, smoking, and sexual practices in the etiology of anal cancer. *Cancer, 101*, 270-280.

Damania, B. (2007). DNA tumor viruses and human cancer. *Trends Microbiol., 15*, 38-44.

Doorbar, J., Ely, S., Sterling, J., McLean, C., & Crawford, L. (1991). Specific interaction between HPV-16 E1-E4 and cytokeratins results in collapse of the epithelial cell intermediate filament network. *Nature, 352*, 824-827.

Doorbar, J. (2005). The papillomavirus life cycle. *J. Clin. Virol., 32* Suppl, S7-S15.

D'Souza, G., Kreimer, A.R., Viscidi, R., Pawlita, M., Fakhry, C., Koch, W.M., Westra, W.H., & Gillison, M.L. (2007). Case-control study of human papillomavirus and oropharyngeal cancer. *N. Engl. J. Med., 356*, 1944-1956.

Duensing, S., & Munger, K. (2004). Mechanisms of genomic instability in human cancer: Insights from studies with human papillomavirus oncoproteins. *Int. J. Cancer, 109*, 157-162.

Dyson, N., Guida, P., Munger, K., Harlow, E. (1992). Homologous sequences in adenovirus E1A and human papillomavirus E7 proteins mediate interaction with the same set of cellular proteins. *J. Virol., 66*, 6893-6902.

El-Omar, E.M., Ng, M.T., & Hold, G.L. (2008). Polymorphisms in Toll-like receptor genes and risk of cancer. *Oncogene, 27*, 244-252.

Espey, D.K., Wu, X.C., Swan, J., Wiggins, C., Jim, M.A., Ward, E., Wingo, P.A., Howe, H.L., Ries, L.A., Miller, B.A., Jemal, A., Ahmed, F., Cobb, N., Kaur, J.S., Edwards, B.K. (2007). Annual Report to the nation on the status of cancer, 1975-2004, featuring cancer in American Indians and Alaska natives. *Cancer, 110*, 2119-2152.

Favre-Bonvin, A., Reynaud, C., Kretz-Remy, C., & Jalinot, P. (2005). Human papillomavirus type 18 E6 protein binds the cellular PDZ protein TIP-2/GIPC, which is involved in transforming growth factor β signaling and triggers its degradation by the proteasome. *J. Virol., 79*, 4229-4237.

Fehrmann, F., & Laimins, L.A. (2003). Human papillomaviruses: Targeting differentiating epithelial cells for malignant transformation. *Oncogene, 22*, 5201-5207.

Ferreux, E., Lont, A.P., Horenblas, S., Gallee, M.P., Raaphorst, F.M., von Knebel Doeberitz, M., Meijder, C.J., Snijders, P.J. (2003). Evidence for at least three alternative mechanisms targeting the p16INK4A/cyclin D/Rb pathway in penile carcinoma, one of which is mediated by high-risk human papillomavirus. *J. Pathol., 201*, 109-118.

Florin, L., Sapp, C., Streeck, R.E., & Sapp, M. (2002). Assembly and translocation of papillomavirus capsid proteins. *J. Virol., 76*, 10009-10014.

Frisch, M., Hjalgrim, H., Jaeger, A.B., & Biggar, R.J. (2000). Changing patterns of tonsillar squamous cell carcinoma in the United States. *Cancer Causes Control, 11*, 489-495.

Funk, J.O., Waga, S., Harry, J.B., Espling, E., Stillman, B., Galloway, D.A. (1997). Inhibition of CDK activity and PCNA-dependent DNA replication by p21 is blocked by interaction with the HPV-16 E7 oncoprotein. *Genes Dev., 11*, 2090-2100.

Garnett, T.O., Filippova, M., & Duerksen-Hughes, P.J. (2006). Accelerated degradation of FADD and procaspase 8 in cells expressing human papilloma virus 16 E6 impair TRAIL-mediated apoptosis. *Cell Death Differ., 13*, 1915-1926.

Gaur, U., & Aggarwal, B.B. (2003). Regulation of proliferation, survival and apoptosis by members of the TNF superfamily. *Biochem. Phamancol.,* 66, 1403-1408.

Gewin, L., Myers, H., Kiyono, T., & Galloway, D.A. (2004). Identification of a novel telomerase repressor that interacts with the human papillomavirus type-16 E6/E6-AP complex. *Genes Dev., 18*, 2269-2282.

Glaunsinger, B.A., Lee, S.S., Thomas, M., Bamks, L., & Javier, R. (2000). Interaction of the PDZ-protein MAGI-1 with adenovirus E4-ORF1 and high-risk papillomavirus E6 oncoproteins. *Oncogene, 19*, 5270-5280.

Gonzalez, S.L., Stremlau, M., He, X., Basile, J.R., & Munger, K. (2001). Degradation of the retinoblastoma tumor suppressor by the human papillomavirus type 16 E7 oncoprotein is important for functional inactivation and its separable from proteasomal degradation of E7. *J. Virol., 75*, 7583-7591.

Gregoire, L., Cubilla, A.L., Reuter, V.E., Haas, G.P., Lancaster, W.D. (1995). Preferential association of human papillomavirus with high-grade histologic variants of penile invasive squamous cell carcinoma. *J. Natl. Cancer Inst., 87*, 1705-1709.

Gross, G., & Pfister, H. (2004). Role of human papillomavirus in penile cancer, penile intraepithelial squamous cell neoplasia and in genital warts. *Med Microbiol. Immunol., 193*, 35-44.

Ha, P.K., & Califano, J.A. (2004). The role of human papillomavirus in oral carcinogenesis. *Crit. Rev. Biol. Med., 15*, 188-196.

Haideman, D.A., Waterboer, T., Pawlita, M., Delis-van Diemen, P., Nindl, I., Leijte, J.A., Bonfrer, J.M., Horenblas, S., Meijer, C.J., & Snijders, P.J. (2007). Human papillomavirus-16 is the predominant type etiologically involved in penile squamous cell carcinoma. *J. Clin. Oncol., 25*, 4550-4556.

Hanahan, D., & Weinberg, R.A. (2000). The hallmarks of cancer. *Cell, 100*, 57-70.

Havard, L., Delvenne, P., Frare, P., Boniver, J. & Giannini, S.L. (2002). Differential production of cytokines and activation of NF-κB in HPV transformed keratinocytes. *Virology, 298*, 271-285.

Havard, L., Rahmouni, S., Boniver, J., & Delvenne, P. (2005). High levels of p105 (NFKB1) and p100 (NFKB2) proteins in HPV 16-transformed keratinocytes: Role of E6 and E7 oncoproteins. *Virology, 331*, 357-366.

Hausen, H. (1999). Viruses in human cancers. *Eur. J. Cancer, 35*, 1174-1181.

Hausen, H. (2000). Papillomavirus causing cancer: Evasion from host-cell control in early events in carcinogenesis. *J. Natl Cancer Inst., 92*, 690-698.

Hausen, H. (2001). Oncogenic DNA viruses. *Oncogene, 20*, 7820-7823.

Hausen, H. (2002). Papillomavirus and cancer: From basic studies to clinical application. *Nat. Rev. Cancer, 2*, 342-350.

Haverkos, H.W. (2004). Virus, chemical and co-carcinogenesis. Oncogene, 23, 6492-6499.

Hebner, C.M., & Laimins, L.A. (2006). Human papillomaviruses: basic mechanisms of pathogenesis and oncogenicity. *Rev. Med. Virol., 16*, 83-97.

Hildesheim, A., & Wang, S.S. (2002). Host and viral genetics and risk of cervical cancer: A review. *Virus Res., 89*, 229-240.

Huibregtse, J.M., Scheffner, M., Howley, P.M. (1993). Localization of the E6-AP regions that direct human papillomavirus E6 binding, association with p53, and ubiquitination of associated proteins. *Mol. Cell Biol., 13*, 4918-4927.

Hussain, S.P., & Harris, C.C. (2007). Inflammation and cancer: An ancient link with novel potentials. *Int. J. Cancer, 121*, 2373-2380.

Hwang, E.S., Nottoli, T., & Dimaio, D. (1995). The HPV16 E5 protein: Expression, detection, and stable complex formation with transmembrane proteins in COS cells. *Virology, 211*, 227-233.

Inoue, J., Gohda, J., Akiyama, T., & Semba, K. (2007). NF-κB activation in development and progression of cancer. *Cancer Sci., 98*, 268-274.

James, M.A., Lee, J.H., Klingelhutz, A.J. (2006). Human papillomavirus type 16 E6 activates NF-κB, induces cIAP-2 expression, and protects against apoptosis in a PDZ binding motif-dependent manner. *J. Virol., 80*, 5301-5307.

Jelen, F., Oleksy, A., Smietana., K., & Otlewski, J. (2003) PDZ domains common players in the cell signaling. *Acta Biochem Pol, 50*, 985-1017.

Johnson, L.G., Madeleine, M.M., Newcomer, L.M., Schwartz, S.M., & Daling, J.R. (2004). Anal cancer incidence and survival: The survellance, epidemiology, and end results experience, 1973-2000. *Cancer, 101*, 281-288.

Jones, D.L., Thompson, D.A., & Munger, K. (1997). Destabilization of the RB tumor suppressor protein and stabilization of p53 contribute to HPV type 16 E7-induced apoptosis. *Virology, 239*, 97-107.

Kanaya, T., Kyo, S., Laimins, L.A. (1997). The 5' region of the human papillomavirus type 31 upstream regulatory region acts as an enhancer which augments virus early expression through the action of YY1. *Virology, 237*, 159-169.

Karin, M., & Greten, F.R. (2005). NF-κB: Linking inflammation and immunity to cancer development and progression. *Nat Rev Immunol., 5*, 749-759.

Karin, M. (2006). Nuclear factor-κB in cancer development and progression. *Nature, 441*, 431-436.

Katzenellenbogen, R.A., Egelkrout, E.M., Vliet-Gregg, P., Gewin, L.C., Gafken, P.R., & Gallowoway, D.A. (2007). NFX1-123 and poly (A) binding proteins synergistically augment activation of telomerase in human papillomavirus type 16 E6-expressing cells. *J. Virol., 81*, 3786-3796.

Khare, S., Pater, M.M., Tang, S.C., Pater, A. (1997). Effect glucocorticoid hormones on viral gene expression, growth, and dysplastic differentiation in HPV 16 –immortalized ectocervical cells. *Exp. Cell Res., 232*, 353-360.

Kiriakidis, S., Andreakos, E., Monaco, C., Foxwell, B., Feldmann, M., & Paleolog, E. (2003). VEGF expression in human macrophages is NF-κB-dependent: Studies using adenoviruses expressing the endogenous NF-κB inhibitor IκBα and a kinase-defective form of the IκB kinase 2. *J. Cell Sci., 116*, 665-674.

Kjellberg, L., Hallmans, G., Ahren, A.M., Johansson, R., Bergman, F., Wadell, G., Angstrom, T., & Dillner, J. (2000). Smoking, diet, pregnancy and oral contraceptive use as risk factors for cervical intra-epithelial neoplasia in relation to human papillomavirus infection. *Br. J. Cancer, 82*, 1332-1338.

Klingelhutz, A.J., Foster, S.A., & McDougall, J.K. (1996). Telomrase activation by the E6 gene product of human papillomavirus type 16. *Nature, 380*, 79-82.

Koskela, P., Anttila, T., Bjorge, T., Brunsvig, A., Dillner, J., Hakama, M., Hakulinen, T., Jellum, E., Lehtinen, M., Lenner, P., Luostarinen, T., Pukkala, E., Saikku, P., Thoresen, S., Youngman, L., & Paavonen, J. (2000). Chlamydia trachomatis infection as a risk factor for invasive cervical cancer. *Int. J. Cancer, 85*, 35-39.

Kreimer, A.R., Clifford, G.M., Boyle, P., & Franceschi, S. (2005). Human papillomavirus types in head and neck squamous cell carcinomas worldwide: A systematic review. 14, 467-475.

Lee, C., & Laimins, L.A. (2004). Role of the PDZ domain-binding motif of the oncoprotein E6 in the pathogenesis of human papillomavirus type 31. *J. Virol., 78*, 12366-12377.

Lefort, K., Mandinova, A., Ostano, P., Kolev, V., Calpini, V., Kolfschoten, I., Devgan, V., Lieb, J., Raffoul, W., Hohl, D., Neel, V., Garlick, J., Chiorino, G., & Dotto, G.P. (2007). Notch1 is a p53 target gene involved in human keratinocyte tumor suppression through negative regulation of ROCK1/2 and MPXKα kinases. *Genes Dev., 21*, 562-577.

Lu, H., Ouyang, W., & Huang, C. (2006). Inflammation, a key event in cancer development. *Mol. Cancer Res.,* 4, 221-233.

Madkan, V.K., Cook-Norris, R.H., Steadman, M.C., Arora, A., Mendoza, N., & Tyring, S.K. (2007). The oncogenic potential of human papillomavirus: A review on the role of host genetics and environmental cofactors. *Br. J. Dermatol., 157*, 228-241.

McCance, D.J., Kalache, A., Ashdown, K., Andrade, L., Menezes, F., Smith, P., Doll, R. (1986). Human papillomavirus types 16 and 18 in carcinomas of the penis from Brazil. *Int. J. Cancer, 37*, 55-59.

Mclntyre, M.C., Ruesch, M.N., Laimins, L.A. (1996). Human papillomavirus E7 oncoproteins bind a single form of cyclin E in a complex with cdk2 and p107. *Vorology, 215*, 73-82.

McMurray, H.R., & McCance, D.J. (2003). Human papillomavirus type 16 E6 activates TERT gene transcription through induction of c-Myc and release of USF-mediated repression. *J. Virol., 77*, 9852-9861.

Mishra, A., Bharti, A.C., Varghese, P., Saluja, D., & Das, B.C. (2006). Differential expression and activation of NF-κB family proteins during oral carcinogenesis: Role of high risk human papillomavirus infection. *Int. J. Cancer, 119*, 2840-2850.

Modis, Y., Trus, B.L., & Harrison, S.C. (2002). Atomic model of the papillomavirus capsid. *EMBO J., 21*, 4754-4762.

Morris, P.J., Dent, C.L., Ring, C.J., Latchman, D.S. (1993a). The octamer binding site in the HPV16 regulatory region produces opposite effects on gene expression in cervical and non-cervical cells. *Nucleic Acids Res., 21*, 1019-1023.

Morris, P.J., Ring, C.J., Lillycrop, K.A., Latchman, D.S. (1993b). Transactivation of the human papilloma virus 16 octamer motif by the octamer binding protein Oct-2 requires both the N and C terminal activation domains. *Nucleic Acids Res., 21*, 4506-4510.

Mueller, N. (1995). Overview: Viral agents and cancer. *Environ. Health Perspect., 103 Suppl. 8*, 259-261.

Munger, K., Basile, J.R., Duensing, S., Eichten, A., Gonzales, S. L., Grace, M., & Zacny, V.L. (2001). Biological activities and molecular targets of the human papillomavirus E7 oncoprotein. *Oncogene, 20*, 7888-7898.

Munger, K., Baldwin, A., Edwards, K.M., Hayakawa, H., Nguyen, C.L., Owens, M., Grace, M., & Huh, K. (2004). Mechanisms of human papillomavirus-induced oncogenesis. *J. Virol., 78*, 11451-11460.

Munoz, N. (2000). Human papillomavirus and cancer: The epidemiological evidence. *J. Clin. Virol., 19*, 1-5.

Narayan, G., Aris-Pulido, H., Koul, S., Vargas, H., Zhang, F.F., Villella, J., Schneider, A., Terry, M.B., Mansukhani, M., & Murty, V.V. (2003). Frequent promoter methylation of CDHI, DAPK, RARB, and HICI genes in carcinoma of cervix uteri: Its relationship to clinical outcome. *Mol. Cancer, 2*, 24.

Narisawa-Saito, M., & Kiyono, T. (2007). Basic mechanisms of high-risk human papillomavirus induced carcinogenesis: Roles of E6 and E7 proteins. *Cancer Sci., 98*, 1505-1511.

Nees, M., Geoghegan, J.M., Hyman, T., Frank, S., Miller, L., Woodworth, C.D. (2001). Papillomavirus type 16 oncogenes downregulate expression of interferon-responsive genes and upregulate proliferation-associated and NF-κB-responsive genes in cervical keratinocytes. *J. Virol., 75*, 4283-4296.

Nguyen, D.X., Westbrook, T.F., McCance, D.J. (2002). Human papillomavirus type 16 E7 maintains elevated levels of the cdc25A tyrosine phosphatase during deregulation of cell cycle arrest. *J. Virol., 76*, 619-632.

Nguyen, M.L., Nguyen, M.M., Lee, D., Griep, A.E., & Lambert, P.F. (2003). The PDZ ligand domain of the human papillomavirus type 16 E6 protein is required for E6's induction of epithelial hyperplasia in vivo. *J. Vorol., 77*, 6957-6964.

Nielson, C.M., Harris, R.B., Dunne, E.F., Abrahamsen, M., Papenfuss, M.R., Flores, R., Markowitz, L.E., & Giuliano, A.R. (2007). Risk factors for anogenital human papillomavirus infection in men. *J. Infect. Dis., 196*, 1137-1145.

Ogden, G.R. (2005). Alcohol and oral cancer. *Alcohol, 35*, 169-173.

Oh, S.T., Kyo, S., & Laimins, L.A. (2001). Telomerase activation by human papillomavirus type 16 E6 protein: Induction of human telomerase reverse transcriptase expression through Myc and GC-rich Sp1 binding sites. *J. Virol., 75*, 5559-5566.

Ohshima, H., Tatemichi, M., & Sawa, T. (2003). Chemical basis of inflammation-induced carcinogenesis. *Arch. Biochem. Biophys., 417*, 3-11.

Ozbun, M.A., & Meyers, C. (1998). Human papillomavirus type 31b E1 and E2 transcript expression correlates with vegetative genome amplification. *Virology, 248*, 218-230.

Parkin, D.M. (2001). Global cancer statistics in the year 2000. *Lancet Oncol., 2*, 533-543.

Parkin, D.M., & Bray, F. (2006). The burden of HPV-related cancers. *Vaccine, 24 Suppl. 3*, S11-S25.

Pascual, A., Pariente, M., Godinez, J.M., Sanchez-Prieto, R., Atienzar, M., Segura, M., & Poblet, E. (2007). High prevalence of human papillomavirus 16 in penile carcinoma. Histol. *Histopathol. 22*, 177-183.

Patel, D., Incassati, A., Wang, N., & McCance, D.J. (2004). Human papillomavirus type 16 E6 and E7 cause polyploidy in human keratinocytes and up-regulation of G2-M-phase proteins. *Cancer Res., 64*, 1299-1306.

Pfister, H. (2003). Human papillomavirus and skin cancer. *J. Natl. Cancer Inst., 31*, 52-56.

Picconi, M.A., Eijan, A.M., Distefano, A.L., Pueyo, S., Alonia, L.V., Gorostidi, S., Teyssie, A.R., Gasabe, A. (2000). Human papillomavirus (HPV) DNA in penile carcinomas in Argentina: Analysis of primary tumors and lymph nodes. *J. Med. Virol., 61*, 65-69.

Pisani, P., Parkin, D.M., Munoz, N., & Ferlay, J. (1997). Cancer and infection: Estimates of the attributable fraction in 1990. Cancer Epidemiol. *Biomarkers Prev., 6*, 387-400.

Psyrri, A. & DiMaio, D. (2008). Human papillomavirus in cervical and head-and-neck cancer. *Nature Clin. Pract. Oncol., 5*, 24-31.

Rubin, M.A., Kleter, B., Zhou, M., Ayala, G., Cubilla, A.L., Quint, W.G., & Pirog, E.C. (2001). Detection and typing of human papillomavirus DNA in penile carcinoma: Evidence for multiple independent pathways of penile carcinogenesis. *Am. J. Pathol., 159*, 1211-1218.

Ruesch, M.N., & Laimins, L.A. (1998). Human papillomavirus oncoproteins alter differentiation-dependent cell cycle exit on suspension in semisolid medium. *Virology, 250*, 19-29.

Shai, A., Brake, T., Somoza, C., & Lambert, P.F. (2007). The human papillomavirus E6 oncogene dysregulates the cell cycle and contributes to cervical carcinogenesis through two independent activities. *Cancer Res., 67*, 1626-1635.

Sano, T., Oyama, T., Kashiwabara, K., Fukuda, T., & Nakajima, T. (1998). Expression status of p16 protein is associated with human papillomavirus oncogenic potential in cervical and genital lesions. (1998). *Am. J. Pathol., 153*, 1741-1748.

Sarkar, F.H., Miles, B.J., Plieth, D.H., Crissman, J.D. (1992). Detection of human papillomavirus in squamous neoplasm of the penis. *J.Urol., 147*, 389-392.

Scheffner, M., Huibregtse, J.M., Vierstra, R.D., & Howley, P.M. (1993). The HPV-16 E6 and E6-AP complex functions as a ubiqitin-protein ligase in the ubiquitination of p53. *Cell, 75*, 495-505.

Schiffman, M., & Castle, P.E. (2003). Human papillomavirus: Epidemiology and public health. *Arch. Pathol. Lab. Med., 127*, 930-934.

Schottenfeld, D., & Beebe-Dimmer, J. (2006). Chronic inflammation: A common and important factor in the pathogenesis of neoplasia. *CA Cancer J. Clin., 56*, 69-83.

Scinicariello, F., Rady, P., Saltzstein, D., Orihuela, E., Tyring, S.K. (1992). Human papillomavirus 16 exhibits a similar integration pattern in primary squamous cell carcinoma of the penis and in its metastasis. *Cancer, 70,* 2143-2148.

Scully, C. (2002). Oral squamous cell carcinoma: From an hypothesis about a virus, to concern about possible sexual transmission. *Oral Oncol., 38*, 227-234.

Senba, M., Kumatori, A., Fujita, S., Jutavijittum, P., Yousukh, A., Moriuchi, T., Nakamura, T., Toriyama, K. (2006). The prevalence of human papillomavirus genotypes in penile cancers from northern Thailand. *J. Med. Virol., 78*, 1341-1346.

Shiboski, C.H., Schmidt, B.L., & Jordan, R.C. (2005). Tongue and tonsil carcinoma: Increasing trends in the U.S. population ages 20-44 years. *Cancer, 103*, 1843-1849.

Sirera, G., Videla, S., Pinol, M., Canadas, M.P., Llatjos, M., Ballesteros, A.L., Garcia-Cuyas, F., Castella, E., Guerola, R., Tural, C., Rey-Joly, C., & Clotet, B. (2006). High prevalence of human papillomavirus infection in the anus, penis and mouth in HIV-positive men. *AIDS, 20*, 1201-1204.

Spitkovsky, D., Hehner, S.P., Hofmann, T.G., Moller, A., & Schmitz, M.L. (2002). The human papillomavirus oncoprotein E7 attenuates NF-κB activation by targeting IκB kinase complex. *J. Biol. Chem., 277*, 25576-25582.

Steenbergen RD, Kramer, D., Braakhuis, B.J., Stern, P.L., Verheijen, R.H., Meijer, C.J., Snijders, P.J. (2004). TSLC1 gene silencing in cervical cancer cell lines and cervical neoplasia. *J. Natl. Cancer Inst., 96*, 294-305.

Storrs, C.H., & Silverstein, S.J. (2007). PATJ, a tight junction-associated PDZ protein, is a novel degradation target of high-risk human papillomavirus E6 and the althernatively spliced isoform 18 E6. *J. Virol., 81*, 4080-4090.

Syrjanen, S. (2003). Human papillomavirus infections and oral tumors. *Med. Microbiol. Immunol., 192*, 123-128.

Syrjanen, S. (2005). Human papillomavirus (HPV) in head and neck cancer. *J. Clin. Virol., 32* Suppl. 1, S59-S66.

Tan, S.H., Leong, L.E., Walker, P.A., & Bernard, H.U. (1994). The human papillomavirus type 16 E2 transcription factor binds with low cooperativity to two flanking sites and represses the E6 promoter through displacement of Sp1 and TFIID. *J. Virol., 68*, 6411-6420.

Thierry, F., Benotmane, M.A., Demeret, C., Mori, M., Teissier, S., & Desaintes, C. (2004). A genomic approach reveals a novel mitotic pathway in papillomavirus carcinogenesis. *Cancer Res., 64*, 895-903.

Thomas, M., Pim, D., & Banks, L. (1999). The role of the E6-p53 interaction in the molecular pathogenesis of HPV. *Oncogene, 18*, 7690-7700.

Tommasino, M., Adamczewski, J.P., Carlotti, F., Barth, C.F., Manetti, R., Contorni, M., Cavalieri, F., Hunt, T., & Crawford, L. (1993). HPV16 E7 protein associates with the protein kinase p33CDK2 and cyclin A. *Oncogene, 8*, 195-202.

Tornesello, M.L., Buonaguro, F.M., Beth-Giraldo, E., Kyalwazi, S.K., & Giraldo, G. (1992). Human papillomavirus (HPV) DNA in penile carcinomas and two cell lines from high-incidence area for genital cancers in Africa. *Int. J. Cancer, 51*, 587-592.

Tornesello, M.L., Duraturo, M.L., Losito, S., Botti, G., Pilott, S., Stefanon, B., De Palo, G., Gallo, A., Buonaguro, L., & Buonaguro, F.M. (2008). Human papillomavirus genotypes and HPV 16 variants in penile carcinoma. *Int. J. cancer, 122*, 132-137.

Tsantoulis, P.K., Kastrinakis, N.G., Tourvas, A.D., Laskaris, G, & Gorgoulis, V.G. (2007). Advances in the biology of oral cancer. *Oral cancer, 43*, 523-534.

Uronis, H.E., & Bendell, J.C. (2008). Anal cancer: An overview. *Oncologist, 12*, 524-534.

Vancurova, I., Wu, R., Miskolci, V., & Sun, S. (2002). Increased p50/p50 NF-κB activation in human papillomavirus type 6- or type 11-induced laryngeal papilloma tissue. *J. Virol., 76*, 1533-1636.

Varma, V.A., Sanchez-Lanier, M., Unger, E.R., Clark, C., Tickman, R., Hewan-Lowe, K., Chenggis, M.L., & Swan, D.C. (1991). Association of human papillomavirus with penile carcinoma. A study using polymerase chain reaction and in situ hybridazation. *Hum. Pathol., 22*, 908-913.

Veldman, T., Liu, X., Yuan, H., & Schlegel, R. (2003). Human papillomavirus E6 and Myc proteins associate in vivo and bind to and cooperatively activate the telomerase reverse transcriptase promoter. *Proc. Natl. Acad. Sci. USA, 100*, 8211-8216.

Villiers, EM, Fauquet, C., Broker, T.R., Bernard, H.U., & Hausen, H. (2004). Classification of papillomavirus. *Virology, 324*, 17-27.

Walboomers, J.M., Jacobs, M.V., Manos, M.M., Boosch, F.X., Kummer, J.A., Shah, K.V., Snijders, P.J., Peto, J., Meijer, C.J., & Munoz, N. (1999). Human papillomavirus is a necessary cause of invasive cervical cancer worldwide. *J. Pathol., 189*, 12-19.

Warnakulasuriya, S., Sutherland, G., & Scully, C. (2005). Tobacco, oral cancer, and treatment of dependence. *Oral Oncol., 41*, 244-260.

Watson, R.A., Thomas, M., Banks, L., & Roberts, S. (2003). Activity of the human papillomavirus E6 PDZ-binding motif correlates with an enhanced morphological transformation of immortalized human keratinocytes. *J. Cell Sci., 116*, 4925-4934.

Wisdschwendter, A., Muller, H.M., Fiegl, H., Ivarsson, L., Wiedemair, A., Muller-Holzner, E., Goebel, G., Marth, C., Widschwendter, M. (2004). DNA methylation in serum and tumors of cervical cancer patients. *Clin. Cancer Res., 10*, 565-571.

Yugawa, T., Handa, K., Narisawa-Saito, M., Ohno, S., Fujita, M., Kiyono, T. (2007). Regulation of Natch1 gene expression by p53 in epithelial cells. *Mol. Cell Biol., 27*, 3732-3742.

Zerfass-Thome, K., Zwerschke, W., Mannhardt, B., Tindle, R., Botz, J.W., Jansen-Durr, P. (1996). Inactivation of the cdk inhibitor p27KIP1 by the human papillomavirus type 16 E7 oncoprotein. *Oncogene, 13*, 2323-2330.

Zhang, A., Maner, S., Betz, R., Angstrom, T., Stendahl, U., Bergman, F., Zetterberg, A., Wallin, K.L. (2002). Genetic alternations in cervical carcinomas: Frequent low-level amplifications of oncogenes are associated with human papillomavirus infection. *Int. J. Cancer, 101*, 427-433.

Zhang, H.S., Postigo, A.A., Dean, D.C. (1999). Active transcriptional repression by the Rb-E2F complex mediates G1 arrest triggered by p16INK4a, TGFβ, and contact inhibition. *Cell, 97*, 53-61.

Zheng, Z.M., & Baker, C.C. (2006). Papillomavirus genome structure, expression, and post-transcriptional regulation. *Front Biosci., 11*, 2286-2302.

In: DNA Tumor Viruses
Editor: H. E. Tao

ISBN 978-1-60692-111-1

Chapter V

PRESENT FOCUS ON HPV VIRUS GENERATED CARCINOMA

Viroj Wiwanitkit
Wiwanitkit House, Bangkhae, Bangkok Thailand 10160.

ABSTRACT

Human papilloma virus (HPV) is a DNA virus that can be the cause of many carcinomas. This DNA virus is proved as DNA tumor virus. The cervical carcinoma is the main carcinoma that relates to HPV infection. Screening for the HPV DNA is the new laboratory analysis in gynecology. Also, the vaccine for HPV is presently available. In this work, the present new knowledge on HPV laboratory diagnosis, therapy and prevention are reviewed and discussed. In addition, the author also performs a metanlysis on the previous reports on cervical carcinoma, the most common female carcinoma in Thailand, and HPV virus infection in Thailand.

INTRODUCTION HPV VIRUS AND CARCINOMA

Human papillomavirus (HPV) is an important double-stranded DNA virus associated with a broad spectrum of clinical states. The virus is transmitted by direct contact with site-subtype specificity. It is accepted as tumor virus [1]. Previous biochemical and serological studies have demonstrated the existence of at least four distinct types of HPVs causing benign skin pathological lesions [2]. These virapplies show strongly no antigenic relationships; their DNAs differ by their sensitivity to restriction endonucleases, and present little, if any, sequence homology, as detected by molecular hybridization using complementary RNAs transcribed in vitro [2]. The recognition of the virapplies is important owing to the different morphology of the pathological lesions generated and their various oncogenic potentials [3]. There are many types of HPV but four main well-realized classical types. HPV1 is majorly responsible for plantar warts, HPV2 for common (hand) warts, HPV3 has been detected both in flat warts and in the variety of epidermodysplasia verruciformis (EV) in which skin pathological lesions are of flat wart type, the course is relatively more benign, and commonly malignant transformation is not to be imagined [3]. HPV4 was up to now found exclusively in the scenarios of EV with prevalent red and red-brownish plaques and hyper- and depigmentations similar to those of pityriasis versicolor. In all scenarios of this variety of EV cancers occured invariably [3]. For HPV1 related diseases, plantar warts are thick, endophytic, hyperkeratotic pathological lesions caused by human papilloma virus [4]. Because they are frequently mistaken for other conditions, they are often misdiagnosed in athletes [4]. The diagnosis of a plantar wart is made by paring down the hypertrophic epithelium until multiple seeds are detected in the dermis representing the thrombosed vessels supplying the wart [4]. For HPV2 related diseases, most people get warts on their hands at some time in their lives [5]. They are most common in childhood, with 4-20% of schoolchildren getting warts at any one time [5]. For HPV3 and HPV4 related disorders, EV is a rare genodermatosis characterized by persistent human papilloma virus infection with possible autosomal recessive inheritance pattern [6]. Clinicsummarizedy, it is characterized by flat wart-like pathological lesions, scaly hypo- and hyperpigmented macules and/or patches, which resemble pityriasis versicolor, and getment of early beginning nonmelanoma cutaneous carcinomas [6]. In addition to simple hyperplasia, HPV is also related to many malignant disorders. The association of HPV-16 and HPV-18 with anogenital malignancy is mentioned at present.

In oncology, HPV is strongly correlated to cervical carcinoma. HPV infection has been implicated in the intraepithelial cervical changes that cause most abnormal Pap smears [7]. To date, more than ten types of HPVs have been identified [7]. Summarized are smsummarized, nongeted, icosahedral DNA virapplies that share a common internal antigen. In scenarios of cervical HPV infection, the koilocytes and dyskeratocytes are the most frequently seen cell types. Most infections are flat aceto-white pathological lesions [7]. Virologicsummarizedy, HPVs are double-stranded DNA viral particle that infect the cutaneous and mucosal epithelium. Infection by specific HPV types has been related to the getment of cervical carcinoma [8]. HPV infects epithelial cells that undergo terminal differentiation and so encode multiple mechanisms to override the normal regulation of differentiation to generate progeny virions [8]. Risk HPVs play an essential function in the pathogenesis of cervical carcinoma. Although infections by these virapplies are very common in the general population, only few result in clinicsummarizedy relevant pathological lesions [9,10]. A strong association of cervical and oral carcinoma with high-risk human papillomavirus (HPV) 16 and 18 infections underlines the importance of the virus in the pathogenesis of these squamous cell carcinomas [9]. Functionsummarizedy high-risk HPV infection leads to carcinogenesis and tumor progression predominantly through the actions of iral oncogenes [9].

HPV is the necessary but probably not sufficient cause of cervical precarcinoma and carcinoma [11]. Secondary exogenous and endogenous factors, HPV cofactors, may lead to the probability of a carcinoma-associated (oncogenic) HPV infection progressing to cervical precarcinoma and carcinoma [11]. For these cofactors to influence the natural history of HPV infection, they must act on cervical tissue to promote viral persistence, progression to precarcinoma or carcinoma given viral persistence, or both [11]. However, HPV only may not be sufficient to induce carcinomas, because even the E6 and E7 oncogenes ORFs of HPV-16 could not transform rat primary cells immediately after infection, but could do so after a long latency. Cervical carcinoma is preventable and it should be a high priority in summarized national carcinoma control programmes. It would seem as though, to be highly effective, various methods warrant joint usage [12]. Tristen and Bergstrom noted that the screening of the whole population in the 35-64 year age groups every 10 years could reduce the incidence of invasive cervical carcinoma by about 55% [13]. Screening for CIN, the curable pathological lesion, is the good preventive method for control of cervical carcinoma. Several alternatives are available for screening for CIN in present medical practice. According to the study of De Vuyst et al [14], the sensitivity for HPV DNA-PCR is significantly higher than other alternative. For the usage of screening in the geting world with limited resource, the concept of laboratory economics should also be focused.

For skin carcinoma, HPV is also documented as an important etiological factor. Basicsummarizedy, the skin is an organ in which direct contact with virapplies, solar ultraviolet (UV) irradiation and increased susceptibility to immune suppression gather to support viral tumorigenesis [15]. Virapplies transform keratinocytes by activation of carcinoma-promoting genes [5]. Viral proteins may directly act as oncogenes that cause cells to proliferate or generate inflammatory responses and cause regeneration of injured cells that eventusummarizedy lead to malignant transformation [15]. Pathophysiologicsummarizedy, HPV infects cutaneous and mucosal epithelia and induces benign and malignant pathological lesions [16]. Non-melanoma skin carcinoma, encompassing basal cell carcinoma and squamous cell carcinoma, is the most common carcinoma in the Caucasian population, and the incidence has increased dramaticsummarizedy worldwide [16]. The relationship between HPV and non-melanoma skin carcinoma is important clinicsummarizedy since non-melanoma skin carcinoma is the most common form of malignancy among fair-skinned populations [17]. It is well set that solar UV irradiation is the main risk factor for geting non-melanoma skin carcinoma, but a pathogenic expression for HPV in the getment of non-melanoma skin carcinoma has also been proposed [17]. Molecularly, high risk mucosal HPV-types encode two major oncoproteins, E6 and E7, which inactivate two suppressor proteins, p53 and pRb respectively, and are sufficient for host-cell immortalization. A polymorphism resulting in either a proline or an arginine at codon 72 might be a relevant risk factor for mucosal HPV-types-associated non-melanoma skin carcinoma [18]. By contrast, E6 of skin HPV-types fails to interact with p53, but prevents infected cells from UV-generated apoptosis leading thus to the propagation of deleterious UV-generated mutations [18]. The antiapoptotic activities of E6 proteins of cutaneous HPV could be relevant to oncogenesis in the interplay with UV exposure [19]. For EV scenarios, EV results from an abnormal susceptibility to infection by specific HPV genotypes (beta-papillomavirapplies) which include the potentisummarizedy oncogenic HPV5 [20]. Of interest, EV-specific HPVs are considered as harmless for the general population [20]. Jenson et al concluded that in normal people, most of the HPV shown to establish a latent infection of the skin, most likely as normal flora residing in hair follicles; however, in patients with various systemic and localized depressions of cell-mediated immunity, some HPV infections shown to be involved in the getment of nonmelanotic skin carcinoma and its precursor pathological lesions in skin, ususummarizedy in sunlight-exposed settings [21]. They also summarized that the HPV might have a function in promoting proliferative pathological lesions of the skin, although their sites of active infection [21].

While much is realized about the natural history of cervical human HPV infection and its consequences, including cervical intraepithelial neoplasia and cervical carcinoma, relatively little is realized about the natural history of anogenital HPV infection and diseases in men [22]. Concerning the carcinoma of penis, HPV is an important etiological factor similar to the case of cervical carcinoma. Basicsummarizedy, infection with high-risk HPV has been associated with intraepithelial neoplasia and carcinomas at various sites of the anogenital tract, including the cervix, vulva, vagina, penis and anus [23]. HPV specificsummarizedy types 16, 18, 31, 33, 35, 39, 45 and 52 have been related to the etiology of anogenital carcinoma [24]. For penis, herpes genitalis and infections caused by HPV are increasing, particularly in young sexusummarizedy active people [25]. Squamous cell carcinoma is the most common malignancy of the penis and the function of oncogenic HPV-types has been also set in squamous cell carcinoma of the penis [25]. The relationship between carcinomas of the penis and cervix uteri is very important in married couples [26]. Genital HPV infections manifest themselves as condylomas [26]. Malek et al said that the combination of the host of carcinogenic factors and currently rampant immunologic disorders will likely lead to an increase in the historicsummarizedy low incidence of squamous cell carcinoma of the penis [27]. Thereforem any men who carry the virus must be identified and treated [26]. Prevention of squamous cell carcinoma of the penis presupposes an identification of risk factors, early determination of summarized pre-carcinomaous pathological lesions and treatment of phimosis [25].

Laboratory Diagnosis and Screening for HPV DNA as Preventive Measure

A variety of new tests for detecting the presence of oncogenic HPV types in biological specimens are available at present [28]. These are relied on the various stages of infection and viral life cycle [28]. There are basicsummarizedy three classes of detectable markers directly got from HPVs: molecular markers relied on determination of nucleic acid sequences, serological markers relied on determination of antibodies against viral proteins, and cellular markers relied on determination of proteins expressed intracellularly, upon either infection or carcinogenesis [28]. Also, there is a classical marker for determination of HPV-related carcinoma, Papanicolaou smear. Since HPVs now comprise many different HPV types, the polymerase chain reaction (PCR) has been the preferred methodology for virus identification and typing on isolated DNA. At present, HPV infection is mainly diagnosed by molecular methods, since reliable serological tools are not available and culture of the virus is not possible [29]. Methods are divided according to its principle to three groups--hybridization based, PCR based and combined both of this principle. The accurate determination and diagnosis of cervical carcinoma and its malignant precursors (collectively referred to as high-grade cervical disease) represents one of the current chsummarizedenges in clinical medicine and cytopathology [30]. HPVDNA testing is a sensitive test for the presence of squamous intraepithelial pathological lesion, with excellent negative predictive value for the absence of squamous intraepithelial pathological lesion [31]. Also, HPVDNA testing might be an effective screening test in the initial evaluation and management of cytological atypical glandular cells [31]. The advent of molecular diagnostics and the use of whole-genome profiling using DNA microarrays promises to yield improved understanding of the disease process with the subsequent getment of more accurate diagnostic procedures based upon these discoveries [30]. It is accepted that PCR primers that target the L1 or E1 region can be unreliable and may miss more advanced disease, whereas those directed at the E6 or E7 regions, which encode oncogenic products, are preferable because 1) the LI/E1 regions, but never the E6/E7 regions, are lost during integration of viral DNA into host genomic DNA, a process that can represent an integral component of progression from infection to tumorigenesis; and 2) the E6/E7 nucleotide sequence exhibits less nucleotide variation [32]. The determination sets the stage for a prospective follow-up study currently being undertaken to evaluate the prognostic relevance of the determination of high-risk HPV E6/E7 oncogene transcripts for the persistence of a high risk HPV infection, and the possible evolution and further getment of a CIN. Future applications of the assay described may include the monitoring of women in studies investigating antiviral treatment or vaccination.

There are some interesting previous reports on the clinical use of HPV oncogene screening in cervical carcinoma. In 2002, Wang-Johanning et al performed a study to determine the potential clinical applies of real-time PCR and reverse transcription-PCR methods for evaluating HPV-16 E6 and E7 oncogene expression [33]. In this work, ThinPrep cervical samples were tested for expression of oncogenes of HPV-16 by real-time PCR or RT-PCR analysis and were compared with determination of expression by conventional PCR and RT-PCR analysis [33]. According to this study, the determination of HPV-16 E6 and E7 expression by real-time RT-PCR or PCR analysis in ThinPrep cervical cytologic specimens may serve as a quick, reliable, and sensitive tool to identify a subset of patients who express HPV-16 oncoproteins [33]. Sasagawa et l set a new assay to detect the E6-E7 DNA of mucosal HPV by a PCR-based method using four pairs of degenerate LCR and E7 primers (LCR-E7 PCR) and compared this assay with the first generation hybrid captured assay (HCA-I) and the MY09/11-PCR method [34]. According to this study, the discordant results were mostly observed in samples with a low-copy number of the HPV genome or with multiple HPV infection [34]. Naucler et al performed an interesting study to compare HPV DNA test and Papanicolaou tests to screen for cervical carcinoma [35]. They said that the addition of an HPV test to the Papanicolaou test to screen women in their mid-30s for cervical carcinoma effectively reduced the incidence of grade 2 or 3 cervical intraepithelial neoplasia or carcinoma detected by subsequent screening examinations [35]. Mayrand et al performed a similar report and found that as compared with Pap testing, HPV testing had greater sensitivity for the determination of cervical intraepithelial neoplasia [36]. Kulasingam et al reported that testing for HPV had higher sensitivity but lower specificity than thin-layer Pap screening [37]. They also reported that in some settings, particularly where screening intervals are long or haphazard, screening for HPV DNA might be a reasonable alternative to cytology-based screening of reproductive-age women [37]. Costa et al found that the potential combinations of pap with cervicography and with hybrid capture test were more sensitive in detecting cervical intraepithelial neoplasia 2 and 3 as compared with pap alone and retained a positive predictive value similar to that of pap [38]. Kuhn et al HPV DNA testing had a sensitivity equivalent to, or better than, that of cytology [39]. Since HPV DNA testing programs might be easier to implement than cytologic screening, HPV testing should be considered for primary cervical carcinoma screening in low-resource settings [39]. In addition, Manos et al proposed that for women with atypical squamous cells of undetermined significance Pap tests, HPV DNA testing of residual specimens collected for routine cervical cytology could help identify those who have underlying high-grade squamous intraepithelial pathological lesions [40]. Bistoletti et al also studied cost-effectiveness of primary cytology and HPV DNA cervical screening in Sweden [41]. They found that population-based, organized cervical cytology screening between ages 32 to 60 was highly cost-efficient for cervical carcinoma prevention [40]. They also noted that if screening intervals were increased to at least 9 years, combined cytology and HPV DNA screening shown to be still more effective and less costly [41].

HPV Therapy at Present

In the few previous years, extensive research has leadd to document the molecular mechanisms of HPV virus persistence and malignant transformation, confirming a direct function of viral proteins in these processes [42]. Previously, major breakthroughs have emerged from immunological studies [42]. Indeed, these studies have paved the way for medical treatment of HPV infections [42]. There are many new immunomodulation drug for treatment of HPV infection. Imiquimod is a new synthetic molecule with potent immune-modifying activities [43]. Formulated in a 5% vanishing cream as Aldara, this self-applied therapy has shown good efficacy and safety in the treatment of external genital and perianal warts caused by human papillomavirus (HPV) infection (Condyloma acuminata) [43]. The molecule does not demonstrate direct antiviral activity, but through induction of cytokines results in immune-based resolution of wart tissue and reduction of viral burden [43]. Imiquimod increases in interferon and tumor necrosis factor in cultures of cells isolated from animal skin, and increases interleukin-8 concentrations in human keratinocyte and fibroblast cultures [44]. Although in vitro studies have shown that imiquimod has no direct antiviral effects, the drug does exhibit antiviral and antitumor effects in vivo through induction of human cytokines [44]. Although imiquimod cream previously became available for the treatment of genital and perianal warts, the topical mechanism of action of imiquimod is not yet fully understood [44]. It is believed that stimulation of local cytokines by imiquimod leads to a reduction in HPV load; to wart regression and to the normalization of keratinocyte proliferation without evidence of scarring [44]. Treatment with topical 5 % imiquimod cream has shown promising results in the treatment of genital warts in immunocompetent individuals [44]. In addition to new antiviral immunomodulation drug, it remains to be investigated if HPV-specific therapeutic vaccines can be geted [45].

HPV VACCINE: NEW THING TO PREVENT INFECTION

Cervical carcinoma is the second leading cause of carcinoma-related death in women worldwide. Approximately 70% of cervical carcinoma is caused by infection with HPV types 16 and 18 and approximately 90% of genital warts are caused by HPV types 6 and 11 [46]. Vaccine for prevention of the HPV infection is the hope for success in control of cervical carcinoma [46]. The chsummarizedenges of the journey from target identification through getment of a prophylactic quadrivalent HPV vaccine have been met in GARDASIL [46,47]. GARDASIL (Merck, Whitehouse Station, NJ) is a non-infectious recombinant, quadrivalent vaccine set from the highly purified virus-like particles (VLPs) of the major capsid proteins of HPV types 6, 11, 16, and 18. GARDASIL is the first vaccine confirmed for use in women aged 9-26 years for the prevention of cervical carcinoma and genital warts, and vulvar and vaginal precarcinomaous pathological lesions [47]. Prophylactic HPV vaccination has demonstrated 100% efficacy in preventing type-specific persistent infections and associated genital dysplasias [48,49]. The Food and Drug Administration has confirmed GARDASIL, the first vaccine geted to prevent cervical carcinoma and precarcinomaous genital pathological lesions and genital warts owing to certain types of HPV [50]. The vaccine is confirmed for use in females ages 9 years to 26 years [50]. Several questions such as the duration of protection, the need getment of for post-exposure vaccination strategies and availability of such vaccine in low-budget countries are open and needed further discussion [51,52].

Table 1. Some reports on HPV infection and cervical carcinoma in Thailand

Authors	Details
Sukvirach et al [64]	This study was performed to determine the prevalence and HPV type in normal and cervical carcinoma tissues of Thai females. Biopsy samples from 105 invasive cervical carcinoma patients (histologicsummarizedy confirmed) and cervical scrapes from 298 normal subjects were collected. Summarized carcinoma patients and normal subjects were those who attended the Out-patient Division of the National Carcinoma Institute. HPV-DNA was assayed by PCR-based method using general and specific probes. HPV-DNA was found in 95 scenarios (90.48 per cent) of cervical carcinoma with 8 "high risk" types of HPV (HPV 16, 18, 31, 33, 35, 39, 52 and 58) being observed in 88 scenarios (83.81 per cent) and HPV X (unrealized HPV) in 7 scenarios (6.67 per cent). In normal cervical smear, HPV-DNA was found in only 28 scenarios (9.4 per cent); 14 scenarios (4.7 per cent) were found positive for "high risk" type HPV (HPV 16, 33, 52 and 58) and the other 14 scenarios (4.7 per cent) were found to be HPV X.
Poonnaniti A, Bhattarakosol P	

[65]	In this study, purified plasmid HPV-16 DNA and HeLa-NA, containing HPV-18 DNA, were amplified by PCR using Ll consensus primers specific for HPV. The amplified product was then analysed by gel electrophoresis (GE) and dot hybridization (DH). The generic oligonucleotide probes (GPs) labelled with Enhanced Chemiluminescence 3' labelling kit (ECL) were used in DH. The sensitivity of PCR reaction after determining by GE was 1 copy per cell for purified plasmid HPV-16 DNA and 20-80 copies per cell for HeLa-DNA while determining by DH was only 0.1 and 2-8 copies per cell, respectively. Thus, the determination of amplified product by DH using enhanced chemiluminescence system improved not only the specificity but also the sensitivity of HPV determination at least 10-fold.

CERVICAL Carcinoma AND HPV INFECTION: EXPERIENCE IN THAILAND

Women in Thailand have a relatively high risk of geting cervical carcinoma [53].Vatanasapt et al noted that an appropriate cervical carcinoma carcinoma screening program can improve the previous prevalence of it and lead to better results of treatment [54]. Swaddiwudhipong et al said that greater efforts should be made to encourage the use of screening among the older women [55]. Promotion of cervical carcinoma screening among Thai adult women, particularly in the rural area, is recommended [55]. There was a previous report on cervical carcinoma screening among the Thai women, who were immigrants in Australia, not local population in Thailand [56]. Similar low rate among the rural female populations in Thailand can be expected. The medical workers should work harder to increase the coverage rate of cervical carcinoma screening for the Thai females. Wiwanitkit summarized and metanalysed the previous reported results from cervical screening programs set by several settings in Thailand [57]. According to the metanalysis, 101,107 scenarios were studied [57,63]. The oversummarized prevalence of cervical carcinoma is 0.38 % (383/101,107) [57]. The oversummarized prevalence of pre-carcinomaous scenarios and other pathological lesions (infection and inflammation) are 2.39 % (2,405/101,107) and 5.30 % orderly (5,358/101,107) [57]. There is no significant association between the prevalent rate and the setting of the study (region) [57]. These results can show the importance of the cervical carcinoma in Thailand [57]. Indeed, the high determination rate of pre-carcinomaous pathological lesion also implies the usefulness of screening program for secondary prevention of cervical carcinoma in Thailand [57]. The health education seems to play an important function in success of the cervical carcinoma screening program for Thai female populations [57].

REFERENCES

[1] Schulz H. Virus tumors. *Dtsch Med Wochenschr*. 1970 Jun 19;95(25):1367-70.

[2] Orth G, Jablonska S, Breitburd F, Favre M, Croissant O. The human papillomavirapplies. *Bull Carcinoma*. 1978;65(2):151-64.

[3] Jablonska S, Orth G, Jarzabek-Chorzelska M, Obalek S, Rzesa G, Croissant O, Favre M. New getments relating to papillomavirapplies. *Hautarzt*. 1979 Aug;30(8):411-7.

[4] Esterowitz D, Greer KE, Cooper PH, Edlich RF. Plantar warts in the athlete. *Am J Emerg Med*. 1995 Jul;13(4):441-3.

[5] Tackling warts on the hands and feet. *Drug Ther Bull*. 1998 Mar;36(3):22-4.

[6] Gül U, Kiliç A, Gönül M, Cakmak SK, Bayis SS. Clinical aspects of epidermodysplasia verruciformis and review of the literature. *Int J Dermatol*. 2007 Oct;46(10):1069-72.

[7] Roy M, Morin C, Casas-Cordero M, Meisels A. Human papillomavirus and cervical pathological lesions. *Clin Obstet Gynecol*. 1983 Dec;26(4):949-67.

[8] Hebner CM, Laimins LA. Human papillomavirapplies: basic mechanisms of pathogenesis and oncogenicity. *Rev Med* Virol. 2006 Mar-Apr;16(2):83-97.

[9] Nair S, Pillai MR. Human papillomavirus and disease mechanisms: relevance to oral and cervical carcinomas. *Oral Dis*. 2005 Nov;11(6):350-9.

[10]Trunk MJ, Wentzensen N, von Knebel Doeberitz M. Molecular pathogenesis of cervical carcinoma and its first steps. *Pathologe*. 2005 Jul;26(4):283-90.

[11]Castle PE. Beyond human papillomavirus: the cervix, exogenous secondary factors, and the getment of cervical precarcinoma and carcinoma. *J Low Genit Tract Dis*. 2004 Jul;8(3):224-30.

[12]Singer A. Cervical carcinoma screening: state of the art. *Baillieres Clin Obstet Gynaecol*. 1995 ;9:39-64.

[13]Tristen C, Bergstrom S. Cervical carcinoma in geting countries. A threat to reproductive health. *Lakartidningen*. 1996;93:3374-6.

[14]De Vuyst H, Calaeys P, Njiru S, Muchiri L, Steyaert S, De Sutter P, Van Marck E, Bwayo J, Temmerman M. Comparison of pap smear, visual inspection with acetic acid, human papilloma virus DNA-PCR testing a cervicography. *Int J Gynaecol Obstet* 2005; 89: 120 – 6.

[15]Molho-Pessach V, Lotem M. Viral carcinogenesis in skin carcinoma. *Curr Probl Dermatol*. 2007;35:39-51.

[16]Nindl I, Gottschling M, Stockfleth E. Human papillomavirapplies and non-melanoma skin carcinoma: basic virology and clinical manifestations. *Dis Markers*. 2007;23(4):247-59.

[17]Akgül B, Cooke JC, Storey A. HPV-associated skin disease. *J Pathol*. 2006 Jan;208(2):165-75.

[18]Aubin F, Humbey O, Guérrini JS, Mougin C, Laurent R. Non-melanoma skin carcinomas and human papillomavirus. *Ann Dermatol Venereol*. 2003 Dec;130(12 Pt 1):1131-8.

[19]Pfister H. Chapter 8: Human papillomavirus and skin carcinoma. *J Natl Carcinoma Inst Mo*nogr. 2003;(31):52-6.

[20]Orth G. Genetics of epidermodysplasia verruciformis: Insights into host defense against papillomavirapplies. *Semin Immunol*. 2006 Dec;18(6):362-74.

[21]Jenson AB, Geyer S, Sundberg JP, Ghim S. Human papillomavirus and skin carcinoma. *J Investig Dermatol Symp Proc*. 2001 Dec;6(3):203-6.

[22]Palefsky JM. HPV infection in men. *Dis Markers*. 2007;23(4):261-72.

[23]Steenbergen RD, de Wilde J, Wilting SM, Brink AA, Snijders PJ, Meijer CJ. HPV-mediated transformation of the anogenital tract. *J Clin Virol*. 2005 Mar;32 Suppl 1:S25-33.

[24]zur Hausen H. The function of papillomavirapplies in anogenital carcinoma. *Scand J Infect Dis Suppl*. 1990;69:107-11.

[25]Buechner SA. Common skin disorders of the penis. *BJU Int*. 2002 Sep;90(5):498-506.

[26]Dutkiewicz S, Witeska A. Human papillomavirus (HPV) infection in premalignancy and carcinoma of the penis. Clinical relevance. *Mater Med Pol*. 1994 Oct-Dec;26(4):139-41

[27]Malek RS, Goellner JR, Smith TF, Espy MJ, Cupp MR. Human papillomavirus infection and intraepithelial, in situ, and invasive carcinoma of penis. *Urology*. 1993 Aug;42(2):159-70.

[28]Pagliusi SR, Garland SM. International standard reagents for HPV determination. *Dis Markers*. 2007;23(4):283-96.

[29]Molijn A, Kleter B, Quint W, van Doorn LJ. Molecular diagnosis of human papillomavirus (HPV) infections. *J Clin Virol*. 2005 Mar;32 Suppl 1:S43-51.

[30]Malinowski DP. Molecular diagnostic assays for cervical neoplasia: emerging markers for the determination of high-grade cervical disease. *Biotechniques*. 2005 Apr;Suppl:17-23.

[31]Irvin W, Evans SR, Andersen W, Jazaeri A, Taylor P, Stoler M, Pastore L, Rice L. The utility of HPV DNA triage in the management of cytological AGC. *Am J Obstet Gynecol*. 2005 Aug;193(2):559-65

[32]Morris BJ. Cervical human papillomavirus screening by PCR: advantages of targeting the E6/E7 region. *Clin Chem Lab Med*. 2005;43(11):1171-7.

[33]Wang-Johanning F, Lu DW, Wang Y, Johnson MR, Johanning GL. Quantitation of human papillomavirus 16 E6 and E7 DNA and RNA in residual material from ThinPrep Papanicolaou tests using real-time polymerase chain reaction analysis. *Carcinoma*. 2002 Apr 15;94(8):2199-210.

[34]Sasagawa T, Minemoto Y, Basha W, Yamazaki H, Nakamura M, Yoshimoto H, Sakaike J, Inoue M. A new PCR-based assay amplifies the E6-E7 genes of most mucosal human papillomavirapplies (HPV). *Virus Res*. 2000 Apr;67(2):127-39.

[35]Naucler P, Ryd W, Törnberg S, Strand A, Wadell G, Elfgren K, Rådberg T, Strander B, Forslund O, Hansson BG, Rylander E, Dillner J. Human papillomavirus and Papanicolaou tests to screen for cervical carcinoma. *N Engl J Med*. 2007 Oct 18;357(16):1589-97.

[36]Mayrand MH, Duarte-Franco E, Rodrigues I, Walter SD, Hanley J, Ferenczy A, Ratnam S, Coutlée F, Franco EL; Canadian Cervical Carcinoma Screening Trial Study Group. Human papillomavirus DNA versus Papanicolaou screening tests for cervical carcinoma. *N Engl J Med*. 2007 Oct 18;357(16):1579-88.

[37]Kulasingam SL, Hughes JP, Kiviat NB, Mao C, Weiss NS, Kuypers JM, Koutsky LA. Evaluation of human papillomavirus testing in primary screening for cervical

abnormalities: comparison of sensitivity, specificity, and frequency of referral. *JAMA*. 2002 Oct 9;288(14):1749-57.

[38]Costa S, Sideri M, Syrjänen K, Terzano P, De Nuzzo M, De Simone P, Cristiani P, Finarelli AC, Bovicelli A, Zamparelli A, Bovicelli L. Combined Pap smear, cervicography and HPV DNA testing in the determination of cervical intraepithelial neoplasia and carcinoma. *Acta Cytol*. 2000 May-Jun;44(3):310-8.

[39]Manos MM, Kinney WK, Hurley LB, Sherman ME, Shieh-Ngai J, Kurman RJ, Ransley JE, Fetterman BJ, Hartinger JS, McIntosh KM, Pawlick GF, Hiatt RA. Identifying women with cervical neoplasia: using human papillomavirus DNA testing for equivocal Papanicolaou results. *JAMA*. 1999 May 5;281(17):1605-10.

[40]Kuhn L, Denny L, Pollack A, Lorincz A, Richart RM, Wright TC. Human papillomavirus DNA testing for cervical carcinoma screening in low-resource settings. *J Natl Carcinoma Inst*. 2000 May 17;92(10):818-25.

[41]Bistoletti P, Sennfält K, Dillner J. Cost-effectiveness of primary cytology and HPV DNA cervical screening. *Int J Carcinoma*. 2008 Jan 15;122(2):372-6.

[42]Menzo S, Marinelli K, Bagnarelli P, Rolla S, Clementi M. Human papillomavirus infections: new perspectives for prevention and treatment. *New Microbiol*. 2007 Jul;30(3):189-212.

[43]Slade HB, Owens ML, Tomai MA, Miller RL. Imiquimod 5% cream (Aldara). *Expert Opin Investig Drugs*. 1998 Mar;7(3):437-49.

[44]Wiwanitkit V. Imiquimod: a new immunomodulatory drug. *Chula Med J* 2004 May; 48(5): 325 – 38.

[45]Müller M, Gissmann L. Prophylactic and therapeutic HPV immunization. *Hautarzt*. 2007 Jun;58(6):483-4, 486-8.

[46]Bryan JT. Geting an HPV vaccine to prevent cervical carcinoma and genital warts. *Vaccine*. 2007 Apr 20;25(16):3001-6.

[47]Shi L, Sings HL, Bryan JT, Wang B, Wang Y, Mach H, Kosinski M, Washabaugh MW, Sitrin R, Barr E. GARDASIL: prophylactic human papillomavirus vaccine getment--from bench top to bed-side. *Clin Pharmacol Ther*. 2007 Feb;81(2):259-64.

[48]New vaccine prevents cervical carcinoma. *FDA Consum*. 2006 Sep-Oct;40(5):37.

[49]Glastetter E, Kaufmann AM. Humoral and cellular immune response in HPV vaccination. *Hautarzt*. 2007 Jun;58(6):493-4, 496-500.

[50]Gissmann L, Müller M. Getment of prophylactic HPV vaccines. *Coll Antropol*. 2007 Apr;31 Suppl 2:113-5.

[51]Kirnbauer R. HPV vaccines. Prophylactic vaccines from virus-like particles. *Hautarzt*. 2007 Jun;58(6):489-92.

[52]Gariglio P, Benitez-Bribiesca L, Berumen J, Alcocer JM, Tamez R, Madrid V. Therapeutic uterine-cervix carcinoma vaccines in humans. *Arch Med Res*. 1998 Winter;29(4):279-84.

[53]Punyaratabandhu P, Supanvanich S, Tirapat C, Podhipak A. Epidemiologic study of risk factors in carcinoma of the cervix uteri in Thai women. *J Med Assoc Thai* 1982;65:231-9

[54]Vatanasapt V, Sriamporn S, Vatanasapt P. Carcinoma control in Thailand. *Jpn J Clin Oncol* 2002 ;32 Suppl:S82-91

[55]Swaddiwudhipong W, Chaovakiratipong C, Nguntra P, Mahasakpan P, Lerdlukanavonge P, Koonchote S. Effect of a mobile unit on changes in knowledge and use of cervical carcinoma screening among rural Thai women. *Int J Epidemiol* 1995;24:493-8

[56]Jirojwong S, Maclennan R, Manderson L. Health beliefs and Pap smears among Thai women in Brisbane, Australia. *Asia Pac J Public Health* 2001;13:20-3

[57]Wiwanitkit V. Screening for cervical carcinoma, results from Thailand. *Asian Pac J Carcinoma Prev*. 2006 Apr-Jun;7(2):329-30.

[58]Nopdonrattakoon L. Results of the cervical carcinoma screening programme at Maesot General Hospital. *Bull Dept Med Serv* 1993; 18: 586-591

[59]Veskijkul S, Kanchanaporn K, Chicharoen S, Wattanakitkrailert S, Srivongpanish S, Wacharapiyanan S, Arrerak K. Cervical carcinoma screening in Ratapum District: cost benefit analysis. *Songklanagarind Med J* 1991; 6: 406-412

[60]Kanwatanasirikul T. Cytologic screening for cervical carcinoma in Northern Thailand. *Bull Chiang Mai Assoc Med Sci* 1980; 14: 23-33

[61]Kanjanavirojkul N, Sripa C. Cervical carcinoma screening in Srinagarind Hospital. *Bull Dept Med Serv* 1995; 20: 349-354

[62]Pongnikorn S. Cervical carcinoma screening in Lampang Regional Carcinoma Centre. *Thai Carcinoma J* 1995 ; 21: 116-122

[63]Khunchairuk T. The results of cervical carcinoma screening in Charoenkrung Pracharak Hospital. *J Charoenkrung Pracharak Hosp* 2003 ; 19: 13-17.

[64]Sukvirach S, Pitakpraiwan P, Kosiyatrakul T, Hankittipongpaisarn P. Determination of human papillomavirus DNA in normal and cervical carcinoma tissues of Thai females. *Thai Carcinoma J.* 1994; 20(3-4): 85-91.

[65]Poonnaniti A, Bhattarakosol P. Improvement of PCR determination of HPV-DNA using enhanced chemiluminescence system and dot hybridization. *J Med Assoc Thai.* 1996 Dec;79 Suppl 1:S96-103.

In: DNA Tumor Viruses
Editor: H. E. Tao
ISBN 978-1-60692-111-1

Chapter VI

HUMAN HERPESVIRUS-8 AND CORTICOSTEROIDS

Celeste Lujan Pérez and Mónica Isabel Tous
ANLIS Dr. Carlos G. Malbrán, Buenos Aires
Department of Virology, National Institute for Infectious Disease
Argentina

ABSTRACT

Kaposi´s sarcoma (KS) associated herpesvirus (KSHV), also termed human herpesvirus 8 (HHV-8) was discovered in 1994 when Chang and colleagues reported the finding of herpesvirus-like sequences in skin lesions from AIDS-KS patients [1]. The virus is present in all KS clinical forms, and it is also associated with another two neoplasic diseases: multicentric Castleman's disease (MCD) and primary effusion lymphoma (PEL). PEL is a rare case of lymphoma first identified as a subset of body-cavity-based lymphomas BCBL [2]. These lymphomas present initially as a lymphomatous effusion located in body cavities without a contiguous tumor mass. PEL cells can be *in vitro* cultured and cell lines PEL- derived are a well-characterized model for a cancer caused by the expression of a few HHV-8 latent proteins [3-5]. Even though most PEL cells are co-infected with HHV-8 and Epstein Baar virus (EBV), some lines, such BC-3, KS-1, BCP-1, CROAP/3, and BCBL-1 contain only HHV-8 episomals [5-9]. Episomes can turn from latent to viral progenie after stimulation with phorbol ester 12- O – tetradenoyl phorbol-13 acetate (TPA), enlarging 16 times the amount of virus [10]. Serum antibodies that are bound to the virus sprouting in the cell are revealed by indirect immunoflourescence assay (IFA). The use of corticosteroids in transplanted patients increases the occurrence and severity of KS [11]. *In vitro,* hydrocortisone induces a viral lytic reactivation in BCBL-1 cell line [12]. In our laboratory, we have examined the ability of dexamethasone (DEX) to induce HHV-8 lytic phase on BCBL-1. We compared the action of DEX with TPA on cell viability, amount of antigen expression, and the performance of IFA slides to detect antibodies against HHV-8 in KS patients, HIV infected individuals and blood donors. Results were analysed by Fisher exact test. Viable cells were counted daily by trypan blue exclusion for six days. Differences between TPA-induced and control cells were found, but no differences were seen in DEX vs. control cells. Daily viral antigen expression within the cells was examined by IFA using HHV-8 positive pool antisera [13]. Slides containing TPA, or DEX-induced cells were made up and cells expressing the HHV-8 lytic pattern were counted by IFA [14]. No differences (p=0.67) were observed among the amount of fluorescent cells in slides with TPA-or DEX-induced cells. Also, no differences were observed between TPA and DEX on fluorescent pattern. In order to investigate if DEX induced cells may be used for HHV-8 serological diagnosis, 86 HHV-8 positive sera from previously described populations were IFA tested and titered in TPA and DEX slides; 54 HHV-8 negative sera were also included [14]. Results fully matched both slides. Our data demonstrate that DEX could be used for studies of HHV-8 lytic cycle activation events and as an alternative to TPA induction to make HHV-8 IFA slides for serologic diagnosis.

INTRODUCTION

Almost two centuries have passed since Moritz Kaposi introduced what now is known as Kaposi's sarcoma (KS). In those days, he reported a rare multifocal pigmented sarcoma occurring in lower extremities of elderly males from Jewish ascendence [15]. Another three clinical KS forms were added to that first described by Kaposi, which is currently known as "classic" KS, mainly observed in men from Mediterranean and Middle Eastern origin. The second clinical, form or "iatrogenic KS", is observed in organ graft patients who are under immunosuppressive therapy. The "endemic KS", more aggressive than classic KS, may involve lymph nodes in addition to skin and often affects children. This clinical type is very common in East and Central Africa [16] where KS cases have been reported for nearly a century, and such a tumor can account for up to 10% of all diagnosed cancers [17,18]. The fourth form is the "epidemic or AIDS associated KS", introduced in the early 1980s in healthy homosexual men at the beginning of the HIV epidemic [19]. Although the incidence of epidemic KS was around 50% in HIV individuals, nowadays it has declined with the success of HAART for HIV treatment [20]. At that moment, it was suspected that there might be another infectious agent besides HIV that caused KS. Some works had related KS to a herpesvirus [21-24], which was finally found in 1994, when Chang et al. reported the presence of herpesvirus-like sequences in skin lesions from AIDS-KS patients [1]. The virus was then called Kaposi´s sarcoma associated herpesvirus (KSHV). It belongs to the family of human herpesviruses (HHV) that includes herpes simplex 1 and 2–oral and genital herpes– also named HHV-1and 2; varicella-zoster chickenpox/shingles (HHV-3); Epstein-Barr virus –mononucleosis (HHV-4); cytomegalovirus or HHV-5; HHV-6; and HHV-7, or roseolavirus. Further analysis made it possible to classify KSHV as a gamma2-herpesvirus, genus radhinovirus, and was termed Human Herpesvirus 8 (HHV-8). Like other herpesvirus, HHV-8 might only be pathogenic when other cofactors are implicated, such as concurrent infection with HIV or in an immunocompromised host. In healthy infected persons, the virus is relatively benign [10].

Herpesviruses have genes that code for proteins involved in establishment of latency and viral replication, for the blocking or modifying host immune defenses, and transitions from latency to lytic growth, among others [25]. Latency is a state that allows the virus to remain in the infected cell, ensuring cell survival, and it is not recognized as infected by the host immune system [3].

Although the four epidemiological forms of KS have different clinical parameters, such as anatomic involvement and aggressiveness of the clinical course, they have HHV-8 infection in common with indistinguishable histopathology [26]. The tumor presents as a highly vascularized neoplasm, with spindle-shaped cells of endothelial origin accompanied by a variable inflammatory mononuclear cell infiltrate. Hemosiderin pigment, extravased red blood between cells, are also common [27].

HHV-8 is also associated with another two neoplasic diseases: primary effusion lymphoma (PEL) and multicentric Castleman's disease (MCD), from which KSHV-positive plasmablastic lymphomas can arise [28]. PEL presents initially as a lymphomatous effusion located in body cavities without a contiguous tumor mass. It was first identified as a subset of body-cavity-based lymphomas (BCBL) [2] and accounts for about 4% of all HIV-associated lymphomas [29]. PEL cells can be *in vitro* cultured and derived cell lines are a well-characterized model for a cancer caused by the activity of a few HHV-8 latent proteins [3-5]. HHV-8 genomes exist in PEL cell lines as mono- or oligoclonal episomes, while most cells exhibit a latent pattern of gene expression and a very small percentage express genes characteristic of the lytic phase [10,30,31]. Even though many PEL cell lines are co-infected with Epstein Barr Virus (EBV) (e.g., BC-1 and BC-2) [4,32], some lines, such as BC-3, KS-1, BCP-1, CROAP/3 and BCBL-1 contain only HHV-8 episomals ranging from 50-150 form copies [5-9]. Episomes can turn from latent to viral progenie after stimulation with butyrate or 12- O – tetradenoyl phorbol-13 acetate (TPA) enlarging the amount of virus particles released [10,33].

Since the discovery of HHV-8 and its association with KS, several efforts have been made to develop accurate serologic assays for the detection of antibodies to HHV-8, for use in diagnosis and screening. Many seroprevalence studies were based on in-house or commercial immunofluorescence assays (IFA) [14,34-36]. Also, enzyme immunoassays (EIAs) were developed using whole viral lysates [37], synthetic peptides [38,39] and recombinant peptide-carrier protein conjugates [40,41]. However, the specificity and sensitivity of these tests are not completely defined due to the lack of a "gold standard" and well-characterized infected and uninfected patients for assay evaluation. Nevertheless, IFA tests made it possible to study the distribution of HHV-8 antibodies among different risk groups for HIV transmission that differ in their propensity to develop KS [42-44]. PEL cells are the source of HHV-8 antigen for making IFA slides which can contain either induced (lytic-IFA) or noninduced cells where latent associated nuclear antigen (LANA) can be observed. Lytic-IFA showed greater sensitivity than LANA-IFA to assess infected persons [10].

In 1999, Hudnall et al. reported that hydrocortisone could directly induce *in vitro* HHV-8 reactivation in BCBL-1 cell line [12]. Nevertheless, Zoeteweij et al. found a minimal reactivation in BC-1, BC-3, and BCBL-1 PEL-derived cell lines [45]. Since the first description of KS after corticosteroid therapy in 1981 by Leung et al. [46], several cases were reported in patients who had not had AIDS or undergone organ transplantation [11,47-57]. The use of corticosteroids in transplanted patients increases the occurence and severity of KS

[58] [59]. In addition, corticosteroids have been reported to substantially exacerbate the severity of KS in patients with pre-existent disease.

The Study

In our laboratory we have examined the ability of dexamethasone (DEX), a cortisol synthetic derivative, to induce the lytic phase in the HHV-8 positive / EBV negative PEL cell line BCBL-1 (NIH AIDS Reagent Program). DEX was used at a concentration of 10 microM, based on a previous report which demonstrated an enhanced cyitomegalovirus infectivity in DEX pretreated cultures [60]. We compared the action of dexamethasone with TPA on cell viability, amount of antigen expression, and the performance of IFA slides prepared with DEX-induced cells to detect antibodies against HHV-8 in KS patients, HIV-infected individuals and blood donors.

Cells were seeded into culture flasks with three different media (2-3 10^5 cells/ ml). The control flask had cells in growth medium (GM), composed of RPMI 1640, 10% (v/v) fetal bovine serum, 2 mM L-glutamine, 10 mM HEPES. The remaining two flasks contained GM additioned with either 10 microM DEX or 20nM TPA. Viable cells were counted daily by trypan blue exclusion for six days. Differences between TPA-induced and control cells were found by Fisher exact test, but no differences were seen in DEX vs. control cells (Figure 1). TPA had a cytotoxic effect double that of DEX on the first day and a dramatic rise at 6th day (Figure 2).

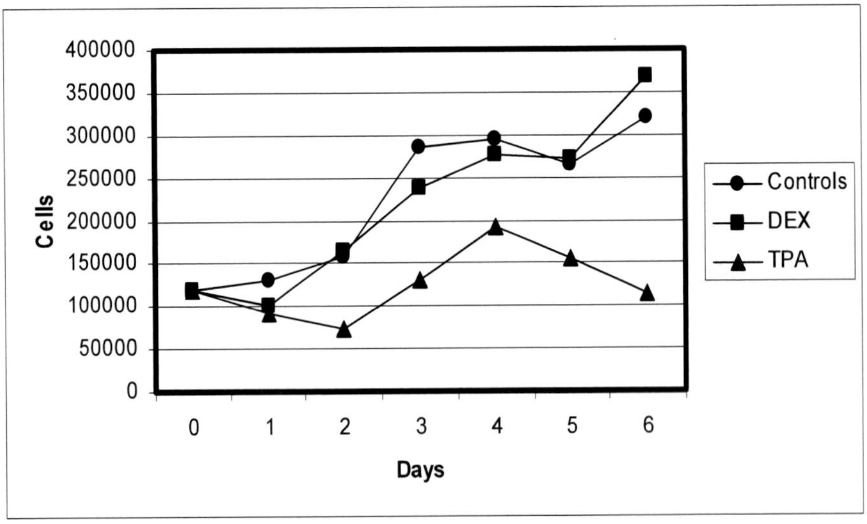

Figure 1. BCBL-1 viable cells (cells / ml) according to treatmet.

Daily viral antigen expression within the cells was examined by indirect immunofluorescence using HHV-8 positive pool antisera [13]. The highest antigen expression was observed at days 3 and 4 after induction, so slides containing TPA or DEX induced cells were made up. Then, cells expressing the specific HHV-8 fluorescent lytic pattern were counted in five fields/well by IFA (400X magnification) using an HHV-8 positive pool antisera. A negative pool antisera and five blood donors were used as a negative IFA control [14]. No differences (p=0.67) were observed among the amount of fluorescent cells in slides made up with TPA or DEX induced cells, by Fisher exact test. Also, no differences were observed between TPA and DEX on fluorescent pattern.

In order to investigate whether DEX-induced cells may be used for HHV-8 serological diagnosis, 86 HHV-8 positive sera from previously described populations [14] were IFA tested and titered in both slides; 54 HHV-8 negative sera were also analysed [14]. Results fully matched both slides.

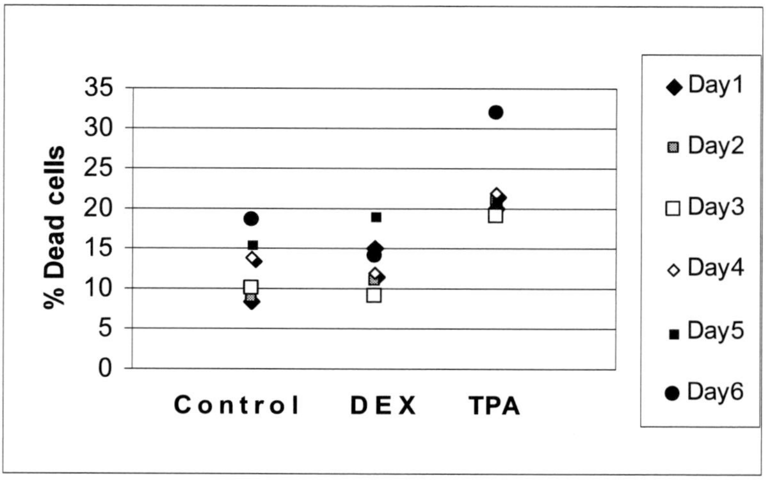

Figure 2. Daily percentage of cell death after TPA or DEX induction.

CONCLUSION

The development of Kaposi sarcoma after corticosteroid therapy had been described many years before the discovery of HHV-8 and its association to KS [46,61-64]. Even though an increased incidence of KS is very common in patients after a solid organ transplant, cases in patients who have not had AIDS or undergone organ transplantation have also been reported. [11,47-55]. In addition, the use of corticosteroids in transplanted patients increases the occurence and severity of KS. As a member of the herpesvirus family, HHV-8 produces opportunistic diseases. The HHV-8 associated malignancies' initiation or excerbation might be due to a secondary effect of drug-induced supression or a depletion of cell-mediated immunity. To address these issues, *in vivo* studies are the most suitable way to find out the mechanism by which HHV-8 diseases can arise, yet systems to investigate PEL in vivo are rare [65]. Since PEL-derived cells were grown *in vitro*, they were considered a good source of HHV-8 antigens, and a useful tool to study viral reactivation under the effect of certain substances such as antivirals and inducing agents [66-69]. Nevertheless, differences between *in vivo* and *in vitro* might be considered because cells are separated from the immunological environment.

Taking into account the background of KS reports after corticosteroid therapy and restricting the action to an *in vitro* environment, controversial results were described. In 1999, Hudnall et al. found that hydrocortisone (HC)1-5 microM acts directly on BCBL-1 cells, activating the lytic cycle of HHV-8 by inducing the accummulation of the linear replicative genomic form of the virus in a time-dependent fashion, increasing mRNA expression of viral genes such as cyclin, Bcl-2, macrophage inflammatory protein-I, interferon regulatory factor-1, and tegument protein by RT-PCR. An increased number of cells expressing lytic antigens was observed by IFA using different HHV-8 positive antisera from infected individuals [12]. However, three years later, Zoeteweij et al. found a minimal viral reactivation only in BCBL-1 cells but not in BC-1or BC-2 PEL cells after treatment with DEX or HC, 30microM each. Besides, only a small increase in the percentage of cells expressing early-lytic PF-8 and late-lytic gpK8.1 proteins was observed in BCBL-1, but not in the other cell lines by flow cytometry. So, they suggest that corticosteroids do not enhance the incidence and severity of KS in transplant recipients by direct cellular effects on KSHV reactivation [45]. Our findings, together with those obtained by Hudnall, provide further evidence for the hypothesis that corticosteroid therapy induces HHV-8 activation by direct effect on infected cells.

DEX has proven to be an inductor as efficient as TPA, and its low cytotoxicity allows a better performance by decreasing dead cell numbers which may produce nonspecific fluorescent patterns. In addition, as oppose to TPA, it is more stable for storage, making more reproducible results among inductions.

Our data demonstrate that Dexamethasone could be used for studies of HHV-8 lytic cycle activation events, and as an alternative to TPA induction to make HHV-8 IFA slides for serologic diagnosis.

REFERENCES

[1] Chang Y, Cesarman E, Pessin MS, Lee F, Culpepper J, Knowles DM, et al. Identification of herpesvirus-like DNA sequences in AIDS-associated Kaposi's sarcoma. *Science.* 1994 Dec 16;266(5192):1865-9.

[2] Cesarman E, Chang Y, Moore PS, Said JW, Knowles DM. Kaposi's sarcoma-associated herpesvirus-like DNA sequences in AIDS-related body-cavity-based lymphomas. *N Engl J Med.* 1995 May 4;332(18):1186-91.

[3] Cesarman E. Kaposi's sarcoma-associated herpesvirus--the high cost of viral survival. *N Engl J Med.* 2003 Sep 18;349(12):1107-9.

[4] Arvanitakis L, Mesri EA, Nador RG, Said JW, Asch AS, Knowles DM, et al. Establishment and characterization of a primary effusion (body cavity-based) lymphoma cell line (BC-3) harboring kaposi's sarcoma-associated herpesvirus (KSHV/HHV-8) in the absence of Epstein-Barr virus. *Blood.* 1996 Oct 1;88(7):2648-54.

[5] Gaidano G, Cechova K, Chang Y, Moore PS, Knowles DM, Dalla-Favera R. Establishment of AIDS-related lymphoma cell lines from lymphomatous effusions. *Leukemia.* 1996 Jul;10(7):1237-40.

[6] Boshoff C, Gao SJ, Healy LE, Matthews S, Thomas AJ, Coignet L, et al. Establishing a KSHV+ cell line (BCP-1) from peripheral blood and characterizing its growth in Nod/SCID mice. *Blood.* 1998 Mar 1;91(5):1671-9.

[7] Carbone A, Cilia AM, Gloghini A, Capello D, Todesco M, Quattrone S, et al. Establishment and characterization of EBV-positive and EBV-negative primary effusion lymphoma cell lines harbouring human herpesvirus type-8. *Br J Haematol.* 1998 Sep;102(4):1081-9.

[8] Said W, Chien K, Takeuchi S, Tasaka T, Asou H, Cho SK, et al. Kaposi's sarcoma-associated herpesvirus (KSHV or HHV8) in primary effusion lymphoma: ultrastructural demonstration of herpesvirus in lymphoma cells. *Blood.* 1996 Jun 15;87(12):4937-43.

[9] Nador RG, Cesarman E, Chadburn A, Dawson DB, Ansari MQ, Sald J, et al. Primary effusion lymphoma: a distinct clinicopathologic entity associated with the Kaposi's sarcoma-associated herpes virus. *Blood.* 1996 Jul 15;88(2):645-56.

[10]Edelman DC. Human herpesvirus 8--a novel human pathogen. *Virol J.* 2005 Sep 2;2:78.

[11]Perez E, Barnadas MA, Garcia-Patos V, Pedro C, Curell R, Sander CA, et al. Kaposi's sarcoma in a patient with erythroblastopenia and thymoma: reactivation after topical corticosteroids. *Dermatology.* 1998;197(3):264-7.

[12]Hudnall SD, Rady PL, Tyring SK, Fish JC. Hydrocortisone activation of human herpesvirus 8 viral DNA replication and gene expression in vitro. *Transplantation.* 1999 Mar 15;67(5):648-52.

[13]Perez C, Tous M, Benetucci J, Gomez J. Correlations between synthetic peptide-based enzyme immunoassays and immunofluorescence assay for detection of human herpesvirus 8 antibodies in different Argentine populations. *J Med Virol.* 2006 Jun;78(6):806-13.

[14]Perez C, Tous M, Gallego S, Zala N, Rabinovich O, Garbiero S, et al. Seroprevalence of human herpesvirus-8 in blood donors from different geographical regions of Argentina, Brazil, and Chile. *J Med Virol.* 2004 Apr;72(4):661-7.

[15]Kaposi M. Idiopatische Multiples Pigmentsarkom der Haut. *Arch Dermatol Syphillis.* 1872; 4:265 - 73.

[16]Ziegler JL, Katongole-Mbidde E. Kaposi's sarcoma in childhood: an analysis of 100 cases from Uganda and relationship to HIV infection. *Int J Cancer.* 1996 Jan 17;65(2):200-3.

[17]Mbulaiteye SM, Pfeiffer RM, Whitby D, Brubaker GR, Shao J, Biggar RJ. Human herpesvirus 8 infection within families in rural Tanzania. *J Infect Dis.* 2003 Jun 1;187(11):1780-5.

[18]Dedicoat M, Newton R. Review of the distribution of Kaposi's sarcoma-associated herpesvirus (KSHV) in Africa in relation to the incidence of Kaposi's sarcoma. *Br J Cancer.* 2003 Jan 13;88(1):1-3.

[19]Friedman-Kien AE. Disseminated Kaposi's sarcoma syndrome in young homosexual men. *J Am Acad Dermatol.* 1981 Oct;5(4):468-71.

[20]Goedert JJ. The epidemiology of acquired immunodeficiency syndrome malignancies. *Semin Oncol.* 2000 Aug;27(4):390-401.

[21]Giraldo G, Beth E, Huang ES. Kaposi's sarcoma and its relationship to cytomegalovirus (CMNV). III. CMV DNA and CMV early antigens in Kaposi's sarcoma. *Int J Cancer*. 1980 Jul 15;26(1):23-9.

[22]Giraldo G, Beth E. [Kaposi's sarcoma: presence of herpes type virus in tissue culture in 5 different cases. A possible new model in the study of viruses associated with human tumors]. *C R Acad Sci Hebd Seances Acad Sci D*. 1972 Jul 10;275(2):289-92.

[23]Giraldo G, Beth E, Coeur P, Vogel CL, Dhru DS. Kaposi's sarcoma: a new model in the search for viruses associated with human malignancies. *J Natl Cancer Inst*. 1972 Dec;49(6):1495-507.

[24]Giraldo G, Beth E, Haguenau F. Herpes-type virus particles in tissue culture of Kaposi's sarcoma from different geographic regions. *J Natl Cancer Inst*. 1972 Dec;49(6):1509-26.

[25]Knipe DM HP. *Fields Virology*. 4th ed. Philadelphia; 2001.

[26]Ablashi DV, Chatlynne LG, Whitman JE, Jr., Cesarman E. Spectrum of Kaposi's sarcoma-associated herpesvirus, or human herpesvirus 8, diseases. *Clin Microbiol Rev*. 2002 Jul;15(3):439-64.

[27]Antman K, Chang Y. Kaposi's sarcoma. *N Engl J Med*. 2000 Apr 6;342(14):1027-38.

[28]Cesarman E, Mesri EA. Kaposi sarcoma-associated herpesvirus and other viruses in human lymphomagenesis. *Curr Top Microbiol Immunol*. 2007;312:263-87.

[29]Chen YB, Rahemtullah A, Hochberg E. Primary effusion lymphoma. *Oncologist*. 2007 May;12(5):569-76.

[30]Cesarman E, Moore PS, Rao PH, Inghirami G, Knowles DM, Chang Y. In vitro establishment and characterization of two acquired immunodeficiency syndrome-related lymphoma cell lines (BC-1 and BC-2) containing Kaposi's sarcoma-associated herpesvirus-like (KSHV) DNA sequences. *Blood*. 1995 Oct 1;86(7):2708-14.

[31]Renne R, Zhong W, Herndier B, McGrath M, Abbey N, Kedes D, et al. Lytic growth of Kaposi's sarcoma-associated herpesvirus (human herpesvirus 8) in culture. *Nat Med*. 1996 Mar;2(3):342-6.

[32]Cesarman E, Knowles DM. The role of Kaposi's sarcoma-associated herpesvirus (KSHV/HHV-8) in lymphoproliferative diseases. *Semin Cancer Biol*. 1999 Jun;9(3):165-74.

[33]Yu Y, Black JB, Goldsmith CS, Browning PJ, Bhalla K, Offermann MK. Induction of human herpesvirus-8 DNA replication and transcription by butyrate and TPA in BCBL-1 cells. *J Gen Virol*. 1999 Jan;80 (Pt 1):83-90.

[34]Kedes DH, Operskalski E, Busch M, Kohn R, Flood J, Ganem D. The seroepidemiology of human herpesvirus 8 (Kaposi's sarcoma-associated herpesvirus): distribution of infection in KS risk groups and evidence for sexual transmission. *Nat Med*. 1996 Aug;2(8):918-24.

[35]Lennette ET, Blackbourn DJ, Levy JA. Antibodies to human herpesvirus type 8 in the general population and in Kaposi's sarcoma patients. *Lancet*. 1996 Sep 28;348(9031):858-61.

[36]Inoue N, Mar EC, Dollard SC, Pau CP, Zheng Q, Pellett PE. New immunofluorescence assays for detection of Human herpesvirus 8-specific antibodies. *Clin Diagn Lab Immunol*. 2000 May;7(3):427-35.

[37]Chatlynne LG, Lapps W, Handy M, Huang YQ, Masood R, Hamilton AS, et al. Detection and titration of human herpesvirus-8-specific antibodies in sera from blood donors, acquired immunodeficiency syndrome patients, and Kaposi's sarcoma patients using a whole virus enzyme-linked immunosorbent assay. *Blood.* 1998 Jul 1;92(1):53-8.

[38]Davis DA, Humphrey RW, Newcomb FM, O'Brien TR, Goedert JJ, Straus SE, et al. Detection of serum antibodies to a Kaposi's sarcoma-associated herpesvirus-specific peptide. *J Infect Dis.* 1997 May;175(5):1071-9.

[39]Pau CP, Lam LL, Spira TJ, Black JB, Stewart JA, Pellett PE, et al. Mapping and serodiagnostic application of a dominant epitope within the human herpesvirus 8 ORF 65-encoded protein. *J Clin Microbiol.* 1998 Jun;36(6):1574-7.

[40]Andre S, Schatz O, Bogner JR, Zeichhardt H, Stoffler-Meilicke M, Jahn HU, et al. Detection of antibodies against viral capsid proteins of human herpesvirus 8 in AIDS-associated Kaposi's sarcoma. *J Mol Med.* 1997 Feb;75(2):145-52.

[41]Smith MS, Bloomer C, Horvat R, Goldstein E, Casparian JM, Chandran B. Detection of human herpesvirus 8 DNA in Kaposi's sarcoma lesions and peripheral blood of human immunodeficiency virus-positive patients and correlation with serologic measurements. *J Infect Dis.* 1997 Jul;176(1):84-93.

[42]Renwick N, Halaby T, Weverling GJ, Dukers NH, Simpson GR, Coutinho RA, et al. Seroconversion for human herpesvirus 8 during HIV infection is highly predictive of Kaposi's sarcoma. *Aids.* 1998 Dec 24;12(18):2481-8.

[43]Martin JN, Ganem DE, Osmond DH, Page-Shafer KA, Macrae D, Kedes DH. Sexual transmission and the natural history of human herpesvirus 8 infection. *N Engl J Med.* 1998 Apr 2;338(14):948-54.

[44]Jacobson LP, Jenkins FJ, Springer G, Munoz A, Shah KV, Phair J, et al. Interaction of human immunodeficiency virus type 1 and human herpesvirus type 8 infections on the incidence of Kaposi's sarcoma. *J Infect Dis.* 2000 Jun;181(6):1940-9.

[45]Zoeteweij JP, Rinderknecht AS, Davis DA, Yarchoan R, Blauvelt A. Minimal reactivation of Kaposi's sarcoma-associated herpesvirus by corticosteroids in latently infected B cell lines. *J Med Virol.* 2002 Mar;66(3):378-83.

[46]Leung F, Fam AG, Osoba D. Kaposi's sarcoma complicating corticosteroid therapy for temporal arteritis. *Am J Med.* 1981 Aug;71(2):320-2.

[47]Trattner A, Hodak E, David M, Neeman A, Sandbank M. Kaposi's sarcoma with visceral involvement after intraarticular and epidural injections of corticosteroids. *J Am Acad Dermatol.* 1993 Nov;29(5 Pt 2):890-4.

[48]Krussmann A, Goerg R, Laberke HG, Weidner FO. [Extensive metastatic Kaposi sarcoma in chronic immune suppressed bronchial asthma]. *Hautarzt.* 1993 Apr;44(4):232-4.

[49]Vandercam B, Lachapelle JM, Janssens P, Tennstedt D, Lambert M. Kaposi's sarcoma during immunosuppressive therapy for atopic dermatitis. *Dermatology.* 1997;194(2):180-2.

[50]Fagone S, Cavaleri A, Camuto M, Iannello S, Belfiore F. [Hyperkeratotic Kaposi sarcoma with leg lymphoedema after prolonged corticosteroid therapy for SLE. Case report and review of the literature]. *Minerva Med.* 2001 Jun;92(3):177-202.

[51]Goedert JJ, Vitale F, Lauria C, Serraino D, Tamburini M, Montella M, et al. Risk factors for classical Kaposi's sarcoma. *J Natl Cancer Inst*. 2002 Nov 20;94(22):1712-8.

[52]Monti M, Mancini LL, Ceriani R, Hendrickx I, Guizzardi M. Iatrogenic Kaposi's sarcoma and HCV infection. *J Eur Acad Dermatol Venereol*. 2004 Mar;18(2):191-3.

[53]Fierro MT, Savoia P, Quaglino P, Novero D, Bernengo MG. Disseminated Kaposi's sarcoma associated with idiopathic CD4+ lymphocytopenia and low dose steroid therapy. *Clin Exp Dermatol*. 2005 Jul;30(4):395-7.

[54]Su CC, Lu JJ, Perng CL, Yu FT, Chiu CH. Evolution of human herpesvirus type 8-associated gastric Kaposi's sarcoma following corticosteroid treatment for asthma. *J Formos Med Assoc*. 2006 Feb;105(2):155-9.

[55]Gonzalez-Sixto B, Conde A, Mayo E, Pardavila R, de la Torre C, Cruces M. [Kaposi sarcoma associated with systemic corticosteroid therapy]. *Actas Dermosifiliogr*. 2007 Oct;98(8):553-5.

[56]Corda L, Benerecetti D, Ungari M, Facchetti F, Radaeli E. Kaposi's disease and sarcoidosis. *Eur Respir J*. 1996 Feb;9(2):383-5.

[57]Rady PL, Hodak E, Yen A, Memar O, Trattner A, Feinmesser M, et al. Detection of human herpesvirus-8 DNA in Kaposi's sarcomas from iatrogenically immunosuppressed patients. *J Am Acad Dermatol*. 1998 Mar;38(3):429-37.

[58]Penn I. The changing pattern of posttransplant malignancies. *Transplant Proc*. 1991 Feb;23(1 Pt 2):1101-3.

[59]Vella JP, Sayegh MH. Maintenance pharmacological immunosuppressive strategies in renal transplantation. *Postgrad Med J*. 1997 Jul;73(861):386-90.

[60]Leonardi GP, Lipson SM. Enhanced detection of cytomegalovirus in shell vial culture following MRC-5 monolayer pretreatment with glucocorticoids. *Zentralbl Bakteriol*. 1992 Jun;277(1):90-9.

[61]Perez-Pascual JJ, Perez-Mur J, Casanova JM, Egido R. [Kaposi sarcoma following treatment of giant cell arteritis with corticoids]. *Med Clin (Barc)*. 1992 Dec 12;99(20):793.

[62]Soria C, Gonzalez-Herrada CM, Garcia-Almagro D, Diaz R, Piris MA, Orradre JL. Kaposi's sarcoma in a patient with temporal arteritis treated with corticosteroid. *J Am Acad Dermatol*. 1991 Jun;24(6 Pt 1):1027-8.

[63]Celentano R, Di Pasquale F, Gregorj M, D'Uva M, De Santis S. [Horton arteritis and Kaposi's sarcoma: description of a case]. *Medicina (Firenze)*. 1989 Oct-Dec;9(4):400-1.

[64]Di Giacomo V, Nigro D, Trocchi G, Meloni F, Verallo P, Bottazzi M, et al. Kaposi's sarcoma following corticosteroid treatment for temporal arteritis--a case report. *Angiology*. 1987 Jan;38(1 Pt 1):56-61.

[65]Staudt MR, Kanan Y, Jeong JH, Papin JF, Hines-Boykin R, Dittmer DP. The tumor microenvironment controls primary effusion lymphoma growth in vivo. *Cancer Res*. 2004 Jul 15;64(14):4790-9.

[66]Zhang YJ, Wang KY, Stein DA, Patel D, Watkins R, Moulton HM, et al. Inhibition of replication and transcription activator and latency-associated nuclear antigen of Kaposi's sarcoma-associated herpesvirus by morpholino oligomers. *Antiviral Res*. 2007 Jan;73(1):12-23.

[67]Zhu W, Burnette A, Dorjsuren D, Roberts PE, Huleihel M, Shoemaker RH, et al. Potent antiviral activity of north-methanocarbathymidine against Kaposi's sarcoma-associated herpesvirus. *Antimicrob Agents Chemother*. 2005 Dec;49(12):4965-73.

[68]Toomey NL, Deyev VV, Wood C, Boise LH, Scott D, Liu LH, et al. Induction of a TRAIL-mediated suicide program by interferon alpha in primary effusion lymphoma. *Oncogene*. 2001 Oct 25;20(48):7029-40.

[69]Medveczky MM, Horvath E, Lund T, Medveczky PG. In vitro antiviral drug sensitivity of the Kaposi's sarcoma-associated herpesvirus. *Aids*. 1997 Sep;11(11):1327-32.

In: DNA Tumor Viruses
Editor: H. E. Tao

ISBN 978-1-60692-111-1

Chapter VII

POLYOMAVIRUS AND CANCER

Romina Bonaventura and María Cecilia Freire
Neurovirosis Division, Department of Virology, National Institute for Infectious Disease, ANLIS Dr. Carlos G. Malbrán, Buenos Aires, Argentina. rbonaventura@anlis.gov.ar

ABSTRACT

The polyomavirus family includes two members, BK virus (BKV) and JC virus (JCV), that naturally infect humans. Besides these, the primate polyomavirus, simian vacuolating virus (SV40), was isolated from humans as a consequence of contaminated polio vaccines in the 1950's in theUnited States. Human polyomaviruses are ubiquitous in the human population worldwide and cause disease only in immunocompromised individuals: progressive multifocal leukoencephalopathy (PML) mainly in AIDS patients and polyomavirus-associated nephropathy (PVAN) in renal transplant patients, being the etiologic agent JCV and BKV, respectively. Many studies indicate that these three polyomaviruses are associated with cancers in humans as their genome sequences and protein expression are detected in several types of human neoplasm. Polyomaviruses rely on cellular enzymes and cofactors for DNA replication and these proteins are expressed in S phase; then JCV, BKV and SV40 encode regulatory proteins (large T antigen, middle T antigen and small t antigen) that drive resting cells into S phase through interactions with cell cycle regulators such as pRb and p53. This aberrant stimulation of the cell cycle is a driving force of oncogenic transformation. Here we describe what is known until now about the molecular mechanisms of these interactions.

INTRODUCTION

Polyomaviruses are small non-enveloped DNA viruses that are widely distributed in mammals (such as humans, monkeys and rodents) and birds all over the world. Each one has a limited host range and does not usually produce productive infections in other species. All polyomaviruses are similar in genomic size and organization. They have a double-stranded circular DNA that encodes for six viral proteins. Transcription is bi-directional with a central point in the *ori* (origin of DNA replication), situated in the regulatory region, which divides the genome in two transcription units: the early region in a counterclockwise direction and the late region in a clockwise direction. The division into early and late regions corresponds to the time after infection when their genes are transcribed. The early proteins are large T antigen (Tag), middle t antigen (mT) and small t antigen (tag), and the late ones are the capsid-proteins VP1, VP2 and VP3, as well as the agnoprotein. The regulatory proteins Tag, mTag, tag and agnoprotein are responsible for stimulation of cellular DNA synthesis and for the establishment and maintenance of cellular transformation [12,24,27,48,62].

Human polyomaviruses are JCV and BKV. They were discovered in 1971 and their names derive from the patients whom were first identified [51,86]. They are ubiquitous in the human population worldwide with rates of seropositivity greater than 80% in adults [50,74,91,99]. They cause an asymptomatic primary infection in children, then establish latency mainly in the urinary tract, among other organs. Viral reactivation can lead to pathology in individuals with impaired cellular immune response.

The primate polyomavirus SV40 was first described in 1960 by Stewart et al. [105]; its natural host is the rhesus monkey. However, between 1955-1963 it was accidentally transferred to nearly one hundred million people in the U.S. via SV40-contaminated polio vaccines, since primary cultures of rhesus monkey cells were used to grow early batches of poliovirus.

The genomes of BKV, JCV and SV40 show high nucleotide sequence homology; BKV and JCV share 75%, and these compared to SV40 display 70% homology in their sequences [47,121]. In spite of their high resemblance, these closely-related polyomaviruses show distinct biological behavior and disease pathogenesis [79].

Human Polyomavirus Related Diseases

JCV

In cases of severe impaired immunity, JCV reactivation occurs and spreads to the central nervous system (CNS), where the virus selectively produces a lytic infection in myelin-producing glial cells, the oligodendrocytes [86,123]. This unique tropism for replication has been shown to be determined by JCV early promoter activity significantly higher in glial cells [55,66,67,107]. JCV replication leads to multiple foci of demyelination in the CNS causing the fatal disease progressive multifocal leukoencephalopathy (PML) [5,17]. PML is fatal in 90% of patients within a year after onset of the symptoms and there is no effective treatment for this disease.

HIV-1-associated immunosuppression is the underlying disease in 85% of the PML cases [9,73]. There is evidence that some interactions between HIV-1 encoded trans-regulatory protein (Tat) and JCV genome can enhance the replication of the latter in glial cells [21,22]. It has been shown that there is high homology between the HIV-1 binding site of Tat (TAR) and the JCV late promoter control region. Studies have demonstrated that Tat transactivation is the responsible for the increased activity of the JCV late promoter in glial cells [22]. This transactivation may not require co-infection of glial cells with these two viruses, since Tat can be secreted by HIV-1 infected cells and then diffuse into neighboring cells [106].

Once a rare disease, PML has become more important since the advent of the HIV/AIDS pandemic. Nowadays, the incidence of PML in these patients is nearly 5% [8].

BKV

BKV reactivation leads to polyomavirus-associated nephropathy (PVAN) in renal transplant patients; this occurs as a consequence of viral replication in tubular cells in kidney. PVAN causes a progressive impairment of renal function and ultimate graft loss. From 10 to 60% of renal transplant recipients present active infection by BKV and 5% suffer from PVAN [11,84,92]. There are risks factors that have been implicated in the development of polyomavirus-associated nephropathy which depend on the patient (serostatus, age, gender), the organ (HLA mismatches, injury from rejection, latent infection) and the virus (strain type, regulatory mutants) [56].

There is no accepted antiviral therapy available, but a reduction in immunosuppression permits recovery of host anti-BKV immune response in order to resolve the infection [94]. For this reason it is important to achieve to the diagnosis of PVAN soon after the beginning of the disease. In this respect it has been recommended to do a screening in these patients by determining viral load in urine and plasma since it has been demonstrated that there is a correlation between viral load and the onset of the disease [57].

BKV is also associated with hemorrhagic cystitis in bone marrow transplant recipients [4,20,112] and ureteric stenosis in renal transplant recipients [25,52,58,75].

Polyomavirus Lifecycle

Viral entry begins with the attachment of the virus to the cell surface. The receptors for JCV and BKV are sialic acids, while for SV40 the receptors are MHC I class molecules [13]. After adsorption, polyomaviruses penetrate into the cell by endocytosis–JCV by clathrin-dependent endocytocis, and BKV and SV40 by caveola-mediated endocytosis [1,42,89]. The viruses are then transported through the cytosol to the nucleus using the cytoskeletal transport machinery [36], where they are uncoated.

In the nucleus, transcription of the early region begins and the early proteins Tag, mTag and tag are produced. These proteins drive the cell into S-phase, which is necessary for viral replication, since polyomaviruses rely on cellular enzymes and cofactors for DNA replication. Tag binds to ori promoting unwinding of the double helix and recruitment of cellular proteins that are required for DNA synthesis, and interacts with several proteins that are implicated in cellular signaling pathways [33,37,41,77]. Following early viral transcription there is a switch to DNA replication and the cycle continues with a second switch to expression of late viral genes mediated by Tag, which is also a strong activator of the late viral promoter [55]. The final products of this activation are the proteins VP1, VP2, VP3 and agnoprotein [23]. The capsid-proteins are then transported to the nucleoplasm for assembly, together with the histone-associated viral DNA. The viral DNA is packaged with histones H2A, H2B, H3 and H4, and creates a mini-chromosome structure. Finally, virions are released either by lytic rupture of the host cell or by transport to the surface in cytoplasmatic vesicles [23].

Molecular Mechanism of Oncogenesis

The proteins that have been found to participate in oncogenesis are large T antigen and, to a lesser extent, middle T antigen and small T antigen.

Tag interactions

The main property of Tag in relation to transformation and oncogenicity is its ability to bind and block the functions of tumor suppressor proteins that regulates cell cycle progression, p53 and pRb family (p105 Rb1, p107 and p130 Rb2) [39,40,54,65].

p53

The Tags of primate polyomaviruses contain at least three domains that influence their oncogenic potential: the LXCXE domain, the J domain and the p53-binding domain [46].

p53 protein acts as a checkpoint in the cell cycle and regulates apoptosis; this serves to eliminate transformed cells, being an important mechanism of defense against cancer [72,113]. Inhibition of p53 allows damaged cells to survive, increasing their probability to transform and acquire immortality. Tag forms a complex with p53 through its p53-binding domain inhibiting and sequestering it in the cytoplasm, then promoting cell proliferation [68,80]. It is believed that a defect in p53 nuclear import could contribute to tumorigenesis [30].

pRB

pRbs, the product of the retinoblastoma tumor suppressor gene, forms a family of nuclear proteins (p105 Rb1, p107 and p130 Rb2) which regulates cell proliferation and cell differentiation through its ability to suppress cell cycle progression acting in the G1 phase [83].

pRb inhibit the activity of the E2F family of transcription factors by sequestering them. The regulation of this pathway is through cyclin-dependent kinases which increasingly phosphorylate pRb during cell cycle progression. This leads to reduce its interactions with E2F transcription factors allowing their release and the expression of E2F-dependent genes such as c-fos and c-myc, whose products are needed for entry into S phase [19,116,117].

pRb has two major domains, A and B. The binding site for LXCXE Tag domain is located in domain B, but domain A is required for domain B to assume an active conformation. The binding site which interacts with the E2F family of transcription factors also depends on the A and B domain. Polyomaviruses Tags bind to underphosphorylated forms of p105 Rb1, p107 and p130 Rb2. Once bound to pRbs-E2F complex, Tag recruits, through its J domain, chaperone Hsc70, and activates its ATPase activity, thus producing a conformational change in the complex that results in the releasing of E2F from p105 Rb1, p107 and p130 Rb2. Finally, E2F leads to the activation of genes involved in S-phase progression [15,118]. Besides this, Tag-dependent activation of E2F leads to the formation of viral DNA initiation complexes (theta structures) that may be recognized by the cell as "unlicensed" DNA replication and a DNA damage-like signal leading to ATM (ataxia telangiectasia mutated) pathway activation. Activated ATM may then potentiate the DNA damage response by phosphorylating and stabilizing E2F. Acting directly or through one or more downstream targets, ATM appears to be responsible for the growth advantage to the virus by prolonging S-phase [29].

Recently it has become apparent that Tag can modulate other signaling proteins besides pRb and p53, such as insulin receptor substrate 1 (IRS-1) and β-catenin.

IRS-1

Tag has been shown to directly bind IRS-1 and cause it to be translocated to the nucleus. This interaction may contribute to the process of malignant transformation in childhood medulloblastomas. The consequences of IRS-1 nuclear translocation may be important in Rad51 trafficking and homologous recombination-directed DNA repair. There is also evidence that interaction with IRS-1 is involved in transformation by SV40 Tag [119].

β-catenin

β-catenin is an integral component of the Wnt signaling pathway whose nuclear import contributes to colorectal carcinogenesis. β-catenin interacts with the central domain of Tag in these tumors leading its translocation into the nucleus where it enhances expression of oncogenes such as c-myc and Cyclin D1 [95]. Evidences demonstrate that in Tag-positive tumor cells, β-catenin is found in the nuclei where Tag is present. Moreover, it has been demonstrated that Tag and wild-type β-catenin can form a complex leading to speculation that the association of these two proteins results in the nuclear entry of β-catenin in neoplastic cells [43].

Other interactions

Finally, Tag may have a role in chromosome damage through its ability to bind topoisomerase I and its helicase activity which may contribute to chromosomal breakage and recombination [108]. Also, Tag is known to disrupt the mitotic checkpoint and attenuate radiation-induced mitotic delay [18]. Chromosomal instability, has been observed characterized by chromosomal breakage, dicentric chromosomes, increasing ploidy, deletions and translocations upon polyomaviruses infection [85,95,109].

tag interactions

The second of the two proteins encoded by the early region of primate polyomaviruses is small t antigen (tag). The N-terminal 82 amino acids are the same as the N terminus of large Tag but the C terminus is a unique domain that is incorporated by alternative splicing of the early region primary transcript; this protein is found in both the nucleus and the cytoplasm.

PP2A

Phosphatases are negative regulators of growth promoting signal transduction pathways. One of them is protein phosphatase 2A (PP2A) which negatively regulates entry into mitosis [64] and also antagonizes ATM and ATR in the DNA damage checkpoint [53,88]. PP2A is bound and inhibited by tag [87]. In this manner, tag is able to activate several pathways that promote cell proliferation including the mitogen-activated protein kinase (MAPK) pathway [103], stress-activated protein kinase (SAPK) pathway [115], and PKC-~/NF-nB. tag contains two conserved cysteine cluster motifs (CXCXXC) in C_terminal domain [90] that are thought to bind the catalytic (C, 36-kDa) and structural (A, 63-kDa) subunits of protein phosphatase 2A (PP2A) [76,97]. The association of tag with AC displaces the cellular regulatory B subunit and results in the inhibition of phosphatase activity [122]. It was shown that the region of tag binding to PP2A is not part of the large Tag [6], suggesting that PP2A binding is a specific function of small tag.

Histone modifications

On the other hand, data have shown that tag induce hyperacetylation at major sites of histones H3 and H4 and downregulate H3ser10 phosphorylation [28]. Minichromosomes of polyomaviruses show an unusually high degree of acetylation of these histones compared to normal cell chromatin [98]; it has been demonstrated that mutants that are defective in cell transformation were not hyperacetylated [7].

Histone modifications play important roles in regulating cellular DNA transcription, replication, and repair. According to this, levels of acetylation and phosphorylation of histones and various modifying enzymes are known to fluctuate during the cell cycle.

It has been shown that tag differentially modulates levels of acetylation and phosphorylation by altering the temporal coordination between virus growth and the virus-driven cell cycle [28]. Finally, a new role for stag has been identified to establish and maintain high levels of viral genomes in cells [44].

mT interactions

PP2A

Middle T antigen (mT) is associated with membranes and underlying cytoskeletal elements. It functions by providing a platform for the assembly of cellular signaling proteins [35,61]. Like stag, it binds to the A and C subunits of PP2A [87,114], thus replacing the B subunit of PP2A and consequently altering its substrate specificity and localization.

PI3Ks

Besides this, mT also binds the class IA phosphatidylinositol 3-kinases (PI3Ks) among other proteins. Recently, it has been described that PI3K pathway activation is essential for mT-mediated transformation. It has been demonstrated that there is a requirement of the p110α isoform of PI3Ks in mT transformation [110].

POLYOMAVIRUSES AND HUMAN CANCER

There are numerous reports that demonstrate the presence of human polyomaviruses' DNA sequence, viral RNA and proteins in tumor tissues obtained from patients with different malignancies [43,63,95,108]. However, there is a lack of evidence to affirm with certainty the causative role of polyomaviruses in developing cancer.

To date, JCV has been found in brain tumors [49] in patients with and without PML; in a case of oligoastrocytoma [93], tumors of glial origin [34], pediatric medulloblastoma [69];,and colon cancer [43,59,70,96].

In the case of BKV, virus DNA was detected in brain tumors [26,32,38]; islet tumors [26] neuroblastomas [45] neoplastic prostate tissues [30] Kaposi's sarcoma [81] urinary tract tumors [82]; neoplastic prostate tissues [30,31] and genital tumors [81,82].

Finally, the transformation of human cells by SV40 was first reported in 1962 [100]; they were found to be semipermissive for SV40 replication. The first report of an association of SV40 with human cancer was made by 1974 when the virus was detected in malignant melanoma metastases [104]. There is a frequent association of SV40 with mesothelioma that has led to the hypothesis that the virus is a cofactor in the development of this tumor [16,102,120]. Besides this, SV40 genome has been detected in brain tumors [2,14], non-Hodgkin's lymphoma [101,111] tumors of the choroid plexus [71] ependymomas [10] osteosarcoma tumors [78] medulloblastomas [60,69] and in a meningioma [3].

CONCLUSION

We have exposed the molecular mechanisms by which polyomaviruses interact with different signalling pathways that are involved in cell cycle progression, demonstrating their potential role in developing cell transformation.

In vitro studies have shown oncogenic ability of polyomaviruses in developing transformation, immortalisation, and deregulation of cell cycle in cultured cells. However, there is controversy about their role in induction of human tumors. Although, the polyomaviruses' presence in human tumor tissue is well reported there are investigations that absolutely failed to detect them, suggesting the need for standardization of the molecular techniques used in this proposal.

The frequency of detection of polyomaviruses in human cancer is low compared to their ubiquitous distribution. In this regard, it is proposed that viral replication leads to transformation through its early proteins. Once this is achieved, there is no longer need of viral sequences to maintain transformation, causing their loss, and, consequently, their low frequency of detection.

Whether polyomaviruses act as a cofactor or have a causal role in developing cancer is a matter of discussion. Further investigation is needed to contribute to a better understanding of the role of polyomaviruses in human cancer.

References

[1] Anderson, H. A., Chen Y., Norkin, L.C. 1996. Bound simian virus 40 translocates to caveolin-enriched membrane domains, and its entry is inhibited by drugs that selectively disrupt caveolae. *Mol.Biol.Cell 7*:1825-34.

[2] Arrington, A. S., Butel J.S. 2001. p. 461-489, *Human Polyomaviruses: Molecular and Clinical Perspective:SV40 and human tumors*. Wiley-Liss, New York.

[3] Arrington, A. S., Moore, M.S., Butel, J.S., 2004. 2004. SV40-positive brain tumor in scientist with risk of laboratory exposure to the virus. *Oncogene 23*:2231-2235.

[4] Arthur, R. R., K. V. Shah, S. J. Baust, G. W. Santos, and R. Saral. 1986. Association of BK viruria with hemorrhagic cystitis in recipients of bone marrow transplants. *N Engl J Med 315*:230-4.

[5] Astrom K., M. E., Richardson E.P. Jr. 1958. Progressive multifocal leukoencephalopathy. *Brain 81*:93-127.

[6] Beck GR Jr, Z. B., Moran E. 1998. *Human Tumor Viruses:Introduction to DNA Tumor Viruses: Adenovirus, Simian Virus 40, and Polyomavirus*. American Society for Microbiology, Washington, DC.

[7] Benjamin, T. 1970. Host range mutants of polyomavirus. *Proc. Natl. Acad. Sci. U.S.A. 67*:394-399.

[8] Berger, J. R. 2003. Progressive multifocal leukoencephalopathyin acquired immunodeficiency syndrome:explaining the high incidence and disproportionate frequency of the illness relative to other immunosuppressive conditions. *J. Neurovirol. 9*:38-41.

[9] Berger, J. R. C. M. 1995. Progressive multifocal leukoencephalopathy: the evolution of a disease once considered rare. *J. Neurovirol. 1*:5-18.

[10] Bergsagel, D. J., Finegold, M.J., Butel, J.S., Kupsky, W.J., Garcea, R.L. 1992. DNA sequences similar to those of simian virus 40 in ependymomas and choroid plexus tumors of childhood. *N. Engl. J. Med. 326*:988-993.

[11] Binet, I., V. Nickeleit, H. H. Hirsch, O. Prince, P. Dalquen, F. Gudat, M. J. Mihatsch, and G. Thiel. 1999. Polyomavirus disease under new immunosuppressive drugs: a cause of renal graft dysfunction and graft loss. *Transplantation 67*:918-22.

[12] Bollag, B., C. Prins, E. L. Snyder, and R. J. Frisque. 2000. Purified JC virus T and T' proteins differentially interact with the retinoblastoma family of tumor suppressor proteins. *Virology 274*:165-78.

[13] Breau W.C., A. W. J., Norkin C. 1992. Class I Major Histocompatibility proteins are an essential component of the Simian Virus 40 receptor. *J. Virol. 66*:2037-45.

[14] Butel, J. S., Jafar S., Wong C.,Arrington A.S., Opekun A.R., Finegold M.J., Adam E. 1999. Evidence of SV40 infections in hospitalized children. *Hum Pathol 30*:1496-1502.

[15] Caracciolo, V., K. Reiss, K. Khalili, G. De Falco, and A. Giordano. 2006. Role of the interaction between large T antigen and Rb family members in the oncogenicity of JC virus. *Oncogene 25*:5294-301.

[16] Carbone, A., Rizzo P., Pass H.I. 1997. Simian virus 40, poliovaccines and human tumors, a review of recent developments. *Oncogene 15*:1877-1888.

[17] Cavanagh J., L. M. 1959. Cerebral demyelination associated with disorders of the reticuloendothelial system. *Lancet 33*:524-529.

[18] Chang T.H., R. F. A., Thompson D.A.,Schlegel R. 1997. Disregulation of mitotic checkpoints and regulatory proteins following acute expression of SV40 large T antigen in diploid human cells. *Oncogene 14:*2383-93.

[19] Chestukhin A., L. L., Rudich K., DeCaprio J.A. 2002. Nucleocytoplasmic Shuttling of p130/RBL2: Novel Regulatory Mechanism. *Mol. Cell. Biol. 22*:453-468.

[20] Childs, R., C. Sanchez, H. Engler, J. Preuss, S. Rosenfeld, C. Dunbar, F. van Rhee, M. Plante, S. Phang, and A. J. Barrett. 1998. High incidence of adeno- and polyomavirus-induced hemorrhagic cystitis in bone marrow allotransplantation for hematological malignancy following T cell depletion and cyclosporine. *Bone Marrow Transplant 22*:889-93.

[21] Chowdhury, K., Taylor J.P., Tada H, Rappaport J., Wong.Staal F., Amini S., Khalili K. 1990. Regulation of the human neurotropic virus promoter by JCV T-antigen and HIV-1 tat protein. *Oncogene 5*:1737-1742.

[22] Chowdhury, M., J. P. Taylor, C. F. Chang, J. Rappaport, and K. Khalili. 1992. Evidence that a sequence similar to TAR is important for induction of the JC virus late promoter by human immunodeficiency virus type 1 Tat. *J Virol 66*:7355-61.

[23] Cole, C. 1996. Polyomaviruses. Lippincott-Raven, Philadelphia.

[24] Cole, C. N. 1996. *Fundamental Virology: Polyomavirinae: The viruses and their replication* (3rd edition), 3rd ed. Lippincott, Williams, Wilkins, Philadelphia.

[25] Coleman, D. V., S. D. Gardner, and A. M. Field. 1973. Human polyomavirus infection in renal allograft recipients. *Br Med J 3*:371-5.

[26] Corallini, A., M. Pagnani, P. Viadana, E. Silini, M. Mottes, G. Milanesi, G. Gerna, R. Vettor, G. Trapella, V. Silvani, and et al. 1987. Association of BK virus with human brain tumors and tumors of pancreatic islets. *Int J Cancer 39*:60-7.

[27] Croul S, O. J., Khalili K. 2003. Brain tumors and polyomaviruses. *J Neurovirol 9*:173-182.

[28] Dahl, J., H. I. Chen, M. George, and T. L. Benjamin. 2007. Polyomavirus small T antigen controls viral chromatin modifications through effects on kinetics of virus growth and cell cycle progression. *J Virol 81*:10064-71.

[29] Dahl, J., J. You, and T. L. Benjamin. 2005. Induction and utilization of an ATM signaling pathway by polyomavirus. *J Virol 79*:13007-17.

[30] Das, D., R. B. Shah, and M. J. Imperiale. 2004. Detection and expression of human BK virus sequences in neoplastic prostate tissues. *Oncogene 23*:7031-46.

[31] Das, D., K. Wojno, and M. J. Imperiale. 2008. BK virus as a cofactor in the etiology of prostate cancer in its early stages. *J Virol 82*:2705-14.

[32] De Mattei, M., F. Martini, A. Corallini, M. Gerosa, K. Scotlandi, P. Carinci, G. Barbanti-Brodano, and M. Tognon. 1995. High incidence of BK virus large-T-antigen-coding sequences in normal human tissues and tumors of different histotypes. *Int J Cancer 61*:756-60.

[33] Dean F.B., B., P.,Murakami, Y., Wobbe, C.R., Weissbach, L.,Hurwitz, J. 1987. Simian virus 40(SV40) DNA replication: SV40 large T antigen unwinds DNA containing the SV40 origin of replication. *Proc. Natl. Acad. Sci. U.S.A. 84*:16-20.

[34] Del Valle, L., J. Gordon, M. Assimakopoulou, S. Enam, J. F. Geddes, J. N. Varakis, C. D. Katsetos, S. Croul, and K. Khalili. 2001. Detection of JC virus DNA sequences and expression of the viral regulatory protein T-antigen in tumors of the central nervous system. *Cancer Res 61*:4287-93.

[35] Dilworth, S. 2002. Polyomavirus middle Tantigen and its role in identifying cancer-related molecules. *Nat. Rev. Cancer 2*:951-956.

[36] Dohner K., S. B. 2005. The role of citoskeleton during viral infection. *Curr. Top. Microbiol. Immunol. 285*:67-108.

[37] Domreiter, I., Hoss, A., Arthur, A.K., Fanning, E. 1990. SV40 T antigen binds directly to the large subunit of purified DNA polymerase alpha. *EMBO J. 9*:3329-3336.

[38] Dorries, K., G. Loeber, and J. Meixensberger. 1987. Association of polyomaviruses JC, SV40, and BK with human brain tumors. *Virology 160*:268-70.

[39] Dyson N, B. R., Friend SH, Gooding LR, Hassell JA, Major EO, Pipas JM, Vandyke T, Harlow E. 1990. Large T antigens of many polyomaviruses are able to form complexes with the retinoblastoma protein. *J Virol. 64*:1353-56.

[40] Dyson, N., K. Buchkovich, P. Whyte, and E. Harlow. 1989. The cellular 107K protein that binds to adenovirus E1A also associates with the large T antigens of SV40 and JC virus. *Cell 58*:249-55.

[41] Eash, S., Manley K., Gasparovic M., Querbes W., Atwood A.J. 2006. The human polyomaviruses. *Cell Mol. Life Sci. 63*:865-876.

[42] Eash, S., W. Querbes, and W. J. Atwood. 2004. Infection of vero cells by BK virus is dependent on caveolae. *J Virol 78*:11583-90.

[43] Enam, S., L. Del Valle, C. Lara, D. D. Gan, C. Ortiz-Hidalgo, J. P. Palazzo, and K. Khalili. 2002. Association of human polyomavirus JCV with colon cancer: evidence for interaction of viral T-antigen and beta-catenin. *Cancer Res 62*:7093-101.

[44] Fahrbach K.M., K. R. B., Rundell K. 2008. Role of SV40 ST Antigen in the Persistent Infection of Mesothelial Cells. *Virology 370*:255-263.

[45] Flaegstad, T., P. A. Andresen, J. I. Johnsen, S. K. Asomani, G. E. Jorgensen, S. Vignarajan, A. Kjuul, P. Kogner, and T. Traavik. 1999. A possible contributory role of BK virus infection in neuroblastoma development. *Cancer Res 59*:1160-3.

[46] Frisque, R. J. 2001. Structure and function of JC virus T' proteins. *J Neurovirol 7*:293-7.

[47] Frisque, R. J., G. L. Bream, and M. T. Cannella. 1984. Human polyomavirus JC virus genome. *J Virol 51*:458-69.

[48] Frisque, R. J., White F.A. 1992. The molecular biology of JC virus, causative agent of progressive multifocal leukoencephalopathy. *Mol. Neurovirol.*:25-158.

[49] Gallia, G. L., J. Gordon, and K. Khalili. 1998. Tumor pathogenesis of human neurotropic JC virus in the CNS. *J Neurovirol 4*:175-81.

[50] Gardner, S. D. 1973. Prevalence in England of antibody to human polyomavirus (B.K.). *Br.Med.J. 1*:77-78.

[51] Gardner, S. D., Field A.M., Coleman D.V. & Hulme B. 1971. New human papovavirus (B.K.) isolated from urine after renal transplantation. *Lancet 1*:1253-1257.

[52] Gardner, S. D., E. F. MacKenzie, C. Smith, and A. A. Porter. 1984. Prospective study of the human polyomaviruses BK and JC and cytomegalovirus in renal transplant recipients. *J Clin Pathol 37*:578-86.

[53] Goodarzi A.A., J. J. C., Douglas P., Young D.,Ye R., Moorhead G.B., Lees-Miller S.P., Khanna K.K. 2004. Autophosphorylation of ataxia-telangiectasia mutated is regulated by protin phospatase 2A. *Embo J 23*:4451-61.

[54] Harris, K. F., J. B. Christensen, and M. J. Imperiale. 1996. BK virus large T antigen: interactions with the retinoblastoma family of tumor suppressor proteins and effects on cellular growth control. *J Virol 70*:2378-86.

[55] Henson, J. W., B. L. Schnitker, T. S. Lee, and J. McAllister. 1995. Cell-specific activation of the glial-specific JC virus early promoter by large T antigen. *J Biol Chem 270*:13240-5.

[56] Hirsch, H. 2003. BK virus replication and disease in solid organ transplant recipient: an update. *Curr. Opin. Organ. Transpl. 8*:262-268.

[57] Hirsch, H. H., D. C. Brennan, C. B. Drachenberg, F. Ginevri, J. Gordon, A. P. Limaye, M. J. Mihatsch, V. Nickeleit, E. Ramos, P. Randhawa, R. Shapiro, J. Steiger, M. Suthanthiran, and J. Trofe. 2005. Polyomavirus-associated nephropathy in renal transplantation: interdisciplinary analyses and recommendations. *Transplantation 79*:1277-86.

[58] Hogan, T. F., E. C. Borden, J. A. McBain, B. L. Padgett, and D. L. Walker. 1980. Human polyomavirus infections with JC virus and BK virus in renal transplant patients. *Ann Intern Med 92*:373-8.

[59] Hori, R., Y. Murai, K. Tsuneyama, H. O. Abdel-Aziz, K. Nomoto, H. Takahashi, C. M. Cheng, T. Kuchina, B. V. Harman, and Y. Takano. 2005. Detection of JC virus DNA sequences in colorectal cancers in Japan. *Virchows Arch 447*:723-30.

[60] Huang, H., Reis, R., Yonekawa, Y., Lopes, J.M., Kleihues, P., Ohgaki, H.,. 1999. Identification in human brain tumors of DNA sequences specific for SV40 large T antigen. *Brain Pathol 9*:33-42.

[61] Ichaso N., D. S. 2001. Cell transformation by the middle T-antigen of polyomavirus. *Oncogene 20*:7908-7916.

[62] Imperiale, M. J. 2001. *Human Polyomaviruses: Molecular and Clinical Perspective: The human polyoma viruses: an overview*. Wiley-Liss, New York.

[63] Imperiale, M. J. 2000. The human polyomaviruses, BKV and JCV: molecular pathogenesis of acute disease and potential role in cancer. *Virology 267*:1-7.

[64] Janssens V., G. J. 2001. Protein phosphatase 2A: a highly regulated family of serine/threonine phpsphatases implicated in cell growth and signalling. *Biochem 353*:417-439.

[65] Kang, S., Folk, W.R. 1992. Lymphotropic papovavirus transforms hamster cells without altering the amount or stability of p53. *Virology 191*:754-764.

[66] Kenney, S., V. Natarajan, D. Strike, G. Khoury, and N. P. Salzman. 1984. JC virus enhancer-promoter active in human brain cells. *Science 226*:1337-9.

[67] Kim H.S., H. J. W., Frisque R.J. 2001. p. 73-126, *Human Polyomaviruses: Molecular and Clinical Perspective: Transcription and replication in the human polyomaviruses.* Wiley-Liss, New York.

[68] Knippschild, U., Oren, M., Deppert, W. 1996. Abrogation of wild-type p53 mediated growth-inhibition by nuclear exclusion. *Oncogene 12*:1755-65.

[69] Krynska, B., L. Del Valle, S. Croul, J. Gordon, C. D. Katsetos, M. Carbone, A. Giordano, and K. Khalili. 1999. Detection of human neurotropic JC virus DNA sequence and expression of the viral oncogenic protein in pediatric medulloblastomas. *Proc Natl Acad Sci U S A 96*:11519-24.

[70] Laghi, L., A. E. Randolph, D. P. Chauhan, G. Marra, E. O. Major, J. V. Neel, and C. R. Boland. 1999. JC virus DNA is present in the mucosa of the human colon and in colorectal cancers. *Proc Natl Acad Sci U S A 96*:7484-9.

[71] Lednicky, J. A., Garcea, R.L., Bergsagel, D.J., Butel, J.S. 1995. Natural simian virus 40 strains are present in human choroid plexus and ependymoma tumors. *Virology 212*.

[72] Levine, A. J. 1997. p53, the cellular gatekeeper for growth and division. *Cell 88*:323-331.

[73] Major, E., Ault G.S. 1995. Progressive multifocal leukoencephalopathy: clinical and laboratory observations on a viral induced demyelinating disease in the inmunodeficient patient. *Curr. Opin. Neurol. 8*.

[74] Major, E. O., Amemiya K., Tornatore C.S., Houff S.A., Berger J.R. 1992. Pathogenesis and molecular biology of progressive multifocal leukoencephalopathy, the JC-virus induced demyelinating disease of the human brain. *Clin.Microbiol.Rev. 5*:49-73.

[75] Marshall, W. F., A. Telenti, J. Proper, A. J. Aksamit, and T. F. Smith. 1991. Survey of urine from transplant recipients for polyomaviruses JC and BK using the polymerase chain reaction. *Mol Cell Probes 5*:125-8.

[76] Mateer S.C., F. S. A., Mumby M.C. 1998. Identification of structural elements involved in the interaction of simian virus 40 small tumor antigen with protein phosphatase 2A. *J. Biol. Chem. 273*:35339-35346.

[77] Melendy, T., Stillman, B. 1993. An interaction between replication protein A and SV40 T antigen appears essential for primosome assembly during SV40 DNA replication. *J. Biol. Chem. 268*:3389-95.

[78] Mendoza, S. M., Konishi, T., Miller, C.W.,. 1998. Integration of SV40 in human osteosarcoma DNA. *Oncogene 17*:2457-2462.

[79] Moens, U., Rekvig P. 2001. p. 359-408, *Human Polyomaviruses: Molecular and Clinical Perspective: Molecular biology of BK virus and clinical and basic aspects of BK virus renal infection.*, vol. 1. Wiley-Liss, New York.

[80] Moll U.M. , O. A. G., Haladay R. , Winkfield B., FRazier M.,Zambetti G. 1996. Cytoplasmic sequestration of wild-type p53 protein impairs the G1 checkpoint after DNA damage. *Mol. Cell. Biol. 16*:1126-37.

[81] Monini, P., A. Rotola, L. de Lellis, A. Corallini, P. Secchiero, A. Albini, R. Benelli, C. Parravicini, G. Barbanti-Brodano, and E. Cassai. 1996. Latent BK virus infection and Kaposi's sarcoma pathogenesis. *Int J Cancer 66*:717-22.

[82] Monini, P., A. Rotola, D. Di Luca, L. De Lellis, E. Chiari, A. Corallini, and E. Cassai. 1995. DNA rearrangements impairing BK virus productive infection in urinary tract tumors. *Virology 214*:273-9.

[83] Mulligan G., J. T. 1998. The retinoblastoma gene family: cousins with overlapping interests. *Trends in Genetics 14*:223-229.

[84] Nickeleit, V., H. H. Hirsch, M. Zeiler, F. Gudat, O. Prince, G. Thiel, and M. J. Mihatsch. 2000. BK-virus nephropathy in renal transplants-tubular necrosis, MHC-class II expression and rejection in a puzzling game. *Nephrol Dial Transplant 15*:324-32.

[85] Niv, Y., A. Goel, and C. R. Boland. 2005. JC virus and colorectal cancer: a possible trigger in the chromosomal instability pathways. *Curr Opin Gastroenterol 21*:85-9.

[86] Padgett, B. L., Walker D.L., ZuRhein G.M., Eckroade R.J. 1971. Cultivation of papova-like virus from human brain with progressive multifocal leucoencephalopathy. *Lancet 1*:1257-1260.

[87] Pallas D.C., S. L. K., Martin B.L., Jaspers S., Miller T.B., Brautigan D.L., Roberts T.M. 1990. Polyoma small and middle T antigens and SV40 small T antigen form a stable complexes with protein phosphatase 2A. *Cell 12*:167-176.

[88] Petersen P., C. D. M., You Z., Hunter T., Walter J.C., Walter G. 2006. Protein phospatase 2A antagonizes ATM and ATR in a Cdk2-and Cdc7-independent DNA damage checkpoint. *Mol Cell Biol 26*:1997-2011.

[89] Pho, M. T., A. Ashok, and W. J. Atwood. 2000. JC virus enters human glial cells by clathrin-dependent receptor-mediated endocytosis. *J Virol 74*:2288-92.

[90] Pipas, J. M. 1992. Common and unique features of T antigen enconded by the polyomavirus group. *J. Virol. 66*:3979-3985.

[91] Portolani, M., A. Marzocchi, G. Barbanti-Brodano, and M. La Placa. 1974. Prevalence in Italy of antibodies to a new human papovavirus (BK virus). *J Med Microbiol 7*:543-6.

[92] Randhawa, P. S., and A. J. Demetris. 2000. Nephropathy due to polyomavirus type BK. *N Engl J Med 342*:1361-3.

[93] Rencic, A., J. Gordon, J. Otte, M. Curtis, A. Kovatich, P. Zoltick, K. Khalili, and D. Andrews. 1996. Detection of JC virus DNA sequence and expression of the viral oncoprotein, tumor antigen, in brain of immunocompetent patient with oligoastrocytoma. *Proc Natl Acad Sci U S A 93*:7352-7.

[94] Reploeg, M. D., G. A. Storch, and D. B. Clifford. 2001. Bk virus: a clinical review. *Clin Infect Dis 33*:191-202.

[95] Ricciardiello, L., M. Baglioni, C. Giovannini, M. Pariali, G. Cenacchi, A. Ripalti, M. P. Landini, H. Sawa, K. Nagashima, R. J. Frisque, A. Goel, C. R. Boland, M. Tognon, E. Roda, and F. Bazzoli. 2003. Induction of chromosomal instability in colonic cells by the human polyomavirus JC virus. *Cancer Res 63*:7256-62.

[96] Ricciardiello, L., D. K. Chang, L. Laghi, A. Goel, C. L. Chang, and C. R. Boland. 2001. Mad-1 is the exclusive JC virus strain present in the human colon, and its transcriptional control region has a deleted 98-base-pair sequence in colon cancer tissues. *J Virol 75*:1996-2001.

[97] Rundell, K., E. O. Major, and M. Lampert. 1981. Association of cellular 56,000- and 32,000-molecular-weight protein with BK virus and polyoma virus t-antigens. *J Virol 37*:1090-3.

[98] Schaffhausen, B., Benjamin T.L. 1976. Deficience in histone acetylation in nontransforming host range mutants of polyomavirus. *Proc. Natl. Acad. Sci. U.S.A. 73*:1092-1096.

[99] Shah, K. V., R. W. Daniel, and R. M. Warszawski. 1973. High prevalence of antibodies to BK virus, an SV40-related papovavirus, in residents of Maryland. *J Infect Dis 128*:784-7.

[100] Shein H.M., E. J. F. 1962. Transformation induced by simian virus 40 in human renal ncell cultures: I. Morphology and growth characteristics. *Proc. Natl. Acad. Sci. U.S.A. 48*:1164-1172.

[101] Shivapurkar, N., Harada K., Reddy J., Scheuermann R.H., Xu Y., McKenna R.W., Milchgrub S.,Kroft S.H.,Feng Z., Gazdar A. F. 2002. Presence of simian virus 40 DNA sequences in human lymphomas. *Lancet 359*:851-852.

[102] Shivapurkar, N., Wiethege, T., Wistuba, I.I., Salomon, E., Milchgrub, S.,, and K. M. Muller, Churg, A., Pass, H., Gazdar, A.F. 1999. Presence of simian virus 40 sequences in malignant mesotheliomas and mesothelial cell proliferations. *J.Cell. Biochem. 76*:181-188.

[103] Sontag E., F. S., Kamibayashi C., Robbins D., Cobb M., Mumby M. 1993. The interaction of SV40 small tumor antigen with protein phospatase 2A stimulates the map kinase pathway and induces cell proliferation. *Cell 75*:887-897.

[104] Soriano, F., Shelburne, C.E., Gokcen, M.,. 1974. Simian virus 40 in a human cancer. *Nature 249*:421-424.

[105] Sweet B.H., H. M. R. 1960. The vacuolating virus, SV40. *Proc. Soc. Exp. Biol. Med. 105*:420-427.

[106] Sweet, T. M., L. Del Valle, and K. Khalili. 2002. Molecular biology and immunoregulation of human neurotropic JC virus in CNS. *J Cell Physiol 191*:249-56.

[107] Tada, H., M. Lashgari, J. Rappaport, and K. Khalili. 1989. Cell type-specific expression of JC virus early promoter is determined by positive and negative regulation. *J Virol 63*:463-6.

[108] Tognon, M., A. Corallini, F. Martini, M. Negrini, and G. Barbanti-Brodano. 2003. Oncogenic transformation by BK virus and association with human tumors. *Oncogene 22*:5192-200.

[109] Trabanelli, C., A. Corallini, R. Gruppioni, A. Sensi, A. Bonfatti, D. Campioni, M. Merlin, N. Calza, L. Possati, and G. Barbanti-Brodano. 1998. Chromosomal aberrations induced by BK virus T antigen in human fibroblasts. *Virology 243*:492-6.

[110] Utermark, T., B. S. Schaffhausen, T. M. Roberts, and J. J. Zhao. 2007. The p110alpha isoform of phosphatidylinositol 3-kinase is essential for polyomavirus middle T antigen-mediated transformation. *J Virol 81*:7069-76.

[111] Vilchez, R. A., Madden, C.R., Kozinetz, C.A., Halvorson, S.J., White, Z.S.,, and J. L. Jorgensen, Finch, C.J., Butel, J.S. 2002. Association between simian virus 40 and non-Hodgkin lymphoma. *Lancet 359*:817-823.

[112] Vogeli, T. A., F. Peinemann, S. Burdach, and R. Ackermann. 1999. Urological treatment and clinical course of BK polyomavirus-associated hemorrhagic cystitis in children after bone marrow transplantation. *Eur Urol 36*:252-7.

[113] Vogelstein, B., Lane, D., Levine, A.J. 2000. Surfing the p53 network. *Nature 408*:307-310.

[114] Walter G., R. R., Slaughter C., Mumby M. 1990. Association of protein phosphatase 2A with polyomavirus medium tumor antigen. *Proc. Natl. Acad. Sci. U.S.A. 87*:2521-2525.

[115] Watanabe G., H. A., Lee R.J., Albanese C., Shu I.W.,Kamezis A.N., Zon L., Kyriakis J., Rundell K., Pestell R.G. 1996. Induction of cyclin D1 by simian virus 40 small tumor antigen. *Proc. Natl. Acad. Sci. U.S.A. 93*:12861-12866.

[116] Weinberg, R. A. 1996. E2F and cell proliferation:a world turned upside down. *Cell 85*:457-459.

[117] Weinberg, R. A. 1995. The retinoblastoma protein and cell cycle control. *Cell 81*:323-330.

[118] White M.K, K. K. 2006. Interaction of retinoblastoma protein family members with large T-antigen of primate polyomaviruses. *Oncogene 25*:5286-5293.

[119] White M.K., G. J., Reiss K., Del Valle L., Croul S.,Giordano A., Darbinyan A., Khalili K. 2005. Human polyomaviruses and brain tumors. *Brain Res. Rev. 50*:69-85.

[120] Wiman, K. G., Klein G. 1997. Human immunodeficiency virus (HIV) and JC virus in acquire immune deficiency syndrome (AIDS) patients with progressive multifocal leukoencephalopathy. *Acta Neuropathol 76*:338-346.

[121] Yang, R. C., and R. Wu. 1979. BK virus DNA: complete nucleotide sequence of a human tumor virus. *Science 206*:456-62.

[122] Yang S.I., L. R. L., Estes R., Rundell K., Walter G., Mumby M.C. 1991. Control of protein phosphatase 2A by simian virus 40 small-t antigen. *Mol. Cell. Biol. 11*:1988-1995.

[123] Zu Rhein, G. M., Chou S.M. 1965. Particles resembling papovaviruses in human cerebral demyelinating disease. *Science 148*:1477-1479.

In: DNA Tumor Viruses
Editor: H. E. Tao

ISBN 978-1-60692-111-1

Chapter VIII

Commentary Macrophage Involvement in the Inflammation-Induced Onset of Epstein-Barr Virus-Related Tumor

***Misuzu Shimakage*[1] *and Haruhiko Sakamoto*[2]**

[1]Department of Pediatrics, National Hospital Organization, Wakayama National Hospital, 1138 Wada, Mihama-cho, Hidaka-gun, Wakayama Prefecture 644-0044, Japan;

[2]Department of Pathology, Faculty of Medicine, Kagawa University, 1750-1, Ikenobe, Miki-cho, Kita-gun, Kagawa Prefecture 761-0793, Japan.

Abstract

Epstein-Barr virus (EBV) is known as a causative agent of Burkitt's lymphoma, nasopharyngeal carcinoma, and about 10% of the cases of stomach carcinoma. We have reported EBV gene expression including EBV lytic infection protein in many other human cancers. We have also identified the expression and replication of EBV genes in normal human macrophages, and abnormal histiocytes in Langerhans' cell histiocytosis (LCH). Furthermore, we revealed EBV expression in macrophages in the lesions of LCH, primary lung lymphoma, uterine cervical carcinoma, renal cell carcinoma, tongue cancer, cutaneous T-cell lymphoma, and thyroid carcinoma. Therefore, tumor-associated macrophages in EBV-related tumors are likely to carry EBV, and they appear to induce EBV lytic infection. Lytic infection of EBV in multiple tissues may provoke a strong inflammatory response, because cell lysis induced by virus replication results in intense immune responses against viral proteins. We reported that childhood LCH expressed high level of EBV lytic infection protein, and that, in one case, the administration of acyclovir resulted in complete remission. Anti-herpesvirus drugs seem to be effective against the disease. We reported that uterine cervical carcinoma expresses EBV genes which participate in transformation, and that carcinoma cells of almost all cases of cervical carcinoma also carried the HPV 16 gene. It is interesting that EBV and HPV may cooperate in the development of cervical carcinoma through the long-term inflammation caused by chronic infection with both viruses. The roles of macrophages not only in inflammation but also in the interaction between viruses or viruses and microbes should be clarified. Through such studies of inflammation involving macrophages, the dynamic mechanism leading to EBV oncogenesis will be elucidated.

ABBREVIATIONS:

EBV,	Epstein-Barr virus;
LCH,	Langerhans' cell histiocytosis;
TAM,	tumor-associated macrophage.

Epstein-Barr virus (EBV) is a ubiquitous DNA tumor virus, infecting almost all adults in the world. On the other hand, EBV has been thought to cause some restricted tumors such as nasopharyngeal carcinoma, which occurs frequently in Chinese; Burkitt's lymphoma, which occurs in children in East Africa; or about 10% of gastric carcinoma cases. It is difficult to explain why the widely-distributed EBV causes endemic tumors in such restricted areas or races. We hypothesized that EBV causes a wider variety of human tumors more frequently than known at present. Using the mRNA *in situ* hybridization method, we have reported that EBV genes are expressed in oral carcinoma [1], meso- and hypopharyngeal carcinoma [2], thyroid carcinoma [3], renal cell carcinoma [4], testicular tumor [5], uterine cervical carcinoma [6,7,8] anaplastic large-cell lymphoma [9,10], cutaneous T-cell lymphoma [11], primary leptomeningeal lymphoma [12], and lymphoma originating in the lung [13]. Other human tumors may be associated with EBV infection, and this needs to be extensively examined. In the tumors mentioned above, the frequency of correlation with EBV was very high (more than 90%) in each disease. Through these studies, we frequently observed the expression of EBV lytic infection protein in the tumor cells as well as infiltrating lymphocytes. This evidence suggests the coexistence of activated tumor cells and lytic lymphocytes in the same tumor tissue. It was generally believed that EBV-related tumor cells express limited genes which operate in tumorigenesis, and do not express lytic proteins. However, Hoshikawa et al. reported evidence for lytic EBV infection in EBV-positive gastric carcinoma [14]. Further, Takasaka et al. observed EBV particles in established human gastric cancer cell lines by electron microscopy [15]. These reports suggest the presence of productive EBV infection in human gastric cancer cells. Therefore, not only EBV-carrying lymphocytes but also tumor cells may produce EBV. These lytic infections might be an intense target of inflammation.

It is well-known that EBV infects macrophages as well as B-lymphocytes, T-lymphocytes and epithelial cells. EBV infection of a macrophage cell line was first described by Revoltella et al. [16]. Furthermore, Savard et al. reported a lytic program of primary human macrophages by EBV [17]. We also showed the expression and replication of EBV genes in normal human macrophages [18], and abnormal histiocytes in Langerhans' cell histiocytosis (LCH) [19,20].

Macrophages are derived from bone marrow promonocytes, which develop into monocytes and extravasate into tissues. There, they differentiate into a specific type of resident tissue macrophage, such as microglial cells in the brain, Kupffer cells in the liver, and Langerhans' cells in the skin. Their functions are to protect the host from microbial infections, to regulate tissue remodeling, and repair injury. Macrophages also form a major component of the inflammatory infiltrate in tumors. Such cells are termed tumor-associated macrophages (TAM) [21]. TAM can kill tumor cells, but they also produce growth factors, angiogenic factors, and proteases which degrade the matrix. Through the action of these macrophage-derived factors, tumor cell proliferation, angiogenesis, tumor invasion, and metastasis are accelerated [22]. We showed the mRNA expression of EBV in macrophages infiltrated into LCH [19,20], primary lung lymphoma [13], uterine cervical carcinoma, tongue cancer, renal cell carcinoma, cutaneous T-cell lymphoma, and thyroid carcinoma [manuscript in preparation]. Therefore, TAM in EBV-related tumors involve EBV-carrying macrophages, and they may also produce EBV. These lytic infections of TAM may also be a target of inflammation.

It has been proposed that inflammation causes cancer [22,23], and that about 15% of the global cancer burden is attributable to infectious agents. Tumor necrosis factor, IL-1α, IL-1β, IL-6, IL-8, IL-18, chemokines, matrix metalloproteinase 9 (MMP-9), vascular endothelial growth factor (VEGF), COX-2 (cyclooxygenase), and 5-LOX (lipoxygenase) have been identified to mediate critical roles in suppression of apoptosis, proliferation, angiogenesis, invasion, and metastasis. The expression of these genes is regulated by the transcription factor NF-kB, which is induced by carcinogens, tumor promoters, carcinogenic viral proteins, and γ-irradiation [24]. The correlation between these inflammatory proteins and EBV has not been sufficiently studied, except for NF-kB. As described above, EBV infects lymphocytes, macrophages, and epithelial cells, and is suspected to produce progeny viruses. Lytic infection of EBV in multiple tissues may provoke a strong inflammatory response, because cell lysis induced by virus replication results in marked immune responses against viral proteins. We reported that childhood LCH, which arises after a short latency period following primary EBV infection, expressed high levels of EBV lytic infection protein, and that, in one case, administration of acyclovir resulted in complete remission [19]. Anti-herpesvirus drugs seem to be effective against the disease. More studies are necessary to confirm the effectiveness of acyclovir-based therapy against EBV-induced tumors including LCH. If inflammation caused by EBV production in the tumor tissue is always as intense as we observed, we can expect that the administration of an anti-herpesvirus drug will be more effective and safer than chemotherapy.

Recently, the suppression of HIV by human herpesvirus 6 [25] or human herpesvirus 7 [26] was reported. Further, it has been reported that latently infected murine gammaherpesvirus 68, which is genetically very similar to EBV, confers resistance against *Listeria monocytogenes* and *Yersinia pestis* in mice [27]. Such virus-virus or virus-microbe interactions might be important when considering oncogenesis due to inflammation caused by viral infection. We reported that EBV genes of *Bam* HIW [6], EBNA2 [7], and EBNA LP [8] are expressed in uterine cervical carcinoma tissue. The frequency of correlation was higher with EBV than with human papillomavirus (HPV) [6]. Almost all cervical carcinoma samples also carried the HPV 16 gene [6], and E6-E7 proteins of HPV16 were reported to induce uncontrollable cell growth [28]. We hypothesize that EBV infection may synergistically act with HPV to cause the development of cancer through the long-term inflammation induced by infiltrating macrophages carrying EBV. Interestingly, a correlation between infiltrating macrophages and the risk of cervical intraepithelial neoplasia was reported [29], although it is unknown whether EBV infection is involved in this process. Moreover, the role of macrophages not only in inflammation but also in the interaction between viruses or viruses and microbes should be clarified. Via these studies on tumor and virus-related inflammation involving macrophages, it may be possible to fully elucidate the dynamic mechanism of EBV oncogenesis.

REFERENCES

[1] Shimakage M, Horii K, Tempaku A, Kakudo K, Shirasaka T, Sasagawa T. Association of Epstein-Barr virus with oral cancers. *Hum Pathol* 2002; 33: 608-614.

[2] Shimakage M, Sasagawa T, Yoshino K, Yutsudo M, Kimura M, Yamamoto N, Yanoma S. Expression of Epstein-Barr virus in mesopharyngeal and hypopharyngeal carcinomas. *Hum Pathol* 1999; 30: 1071-1076.

[3] Shimakage M, Kawahara K, Sasagawa T, Inoue H, Yutsudo M, Yoshida A, Yanoma S. Expression of Epstein-Barr virus in thyroid carcinoma correlates with tumor progression. *Hum Pathol* 2003; 34: 1170-1177.

[4] Shimakage M, Kawahara K, Harada S, Sasagawa T, Shinka T, Oka T. Expression of Epstein-Barr virus in renal cell carcinoma. *Oncol Rep* 2007; 18: 41-46.

[5] Shimakage M, Oka T, Shinka T, Kurata A, Sasagawa T, Yutsudo M. Involvement of Epstein-Barr virus expression in testicular tumors. *J Urol* 1996; 156: 253-257.

[6] Sasagawa T, Shimakage M, Nakamura M, Sakaike J, Ishikawa H, Inoue M. Epstein-Barr virus (EBV) genes expression in cervical intraepithelial neoplasm and invasive cervical cancer: a comparative study with human papillomavirus (HPV) infection. *Hum Pahtol* 2000; 31: 318-326.

[7] Shimakage M, Sasagawa T. Detection of Epstein-Barr virus-determined nuclear antigen-2 mRNA by in situ hybridization. *J Virol Methods* 2001; 93: 23-32.

[8] Shimakage M, Harada S, Kawahara K, Oka T, Yanoma S, Horii K, Sasagawa T. (2006) Detection of Epstein-Barr virus nuclear antigen leader protein expression in various human cancers. In C.S. Umar (Ed) *New Developments in Epstein-Barr virus research* (pp261-276) New York, Nova Science Publishers.

[9] Shimakage M, Dezawa T, Tamura S, Tabata T, Aoyagi N, Koike M, Inoue H, Yutsudo M, Hakura A, Ikegami N. A Ki-1-positive cell line expressing Epstein-Barr virus antigen established from a child with Ki-positive lymphoma. *Intervirol* 1993; 36: 215-224.

[10] Shimakage M, Nakamine H, Tamura S, Takenaka T, Yutsudo M, Hakura A. Detection of Epstein-Barr virus transcripts in anaplastic large-cell lymphomas by mRNA in situ hybridization. *Hum Pathol* 1997; 28: 1415-1419.

[11] Shimakage M, Sasagawa T, Kawahara K, Yutsudo M, Kusuoka H, Kozuka T. Expression of Epstein-Barr virus in cutaneous T-cell lymphoma including mycosis fungoides. *Int J Cancer* 2001; 92: 226-231.

[12]Nakajima H, Shimakage M, Takeda Y, Furutama D, Sugino M, Kimura F, Shibayama Y, Hanafusa T. Epstein-Barr virus-associated primary leptomeningeal lymphoma. *Eur J Neurol* 2006; 13: e4-e6.

[13]Shimakage M, Sakamoto H, Harada S, Sasagawa T, Kodama K. Expression of the Epstein-Barr virus in lymphoploliferative diseases of the lung. *Oncol Rep* 2007; 17: 1347-1352

[14]Hoshikawa Y, Satoh Y, Murakami M, Maeta M, Kaibara N, Ito H, Kurata T, Sairenji T. Evidence of lytic infection of Epstein-barr virus (EBV) in EBV-positive gastric carinoma. *J Med Virol* 2002; 99: 351-359. .

[15]Takasaka N, Tajima M, Okinaga K, Satoh Y, Hoshikawa Y, Katsumoto T, Kurata T, Sairenji T. Productive infection of Epstein-Barr virus in EBV-genome-positive epithelial cell lines (GT38, GT39) derived from gastric tissues. *Virology* 1998; 247: 152-159.

[16]Revoltella RP, Vigneti E, Fruscalzo A, Park M, Ragona G, Rocchi G, Calef E. Epstein-barr virus DNA sequences in precursor monocyte-macrophage cell line established from the bone marrow of children with maturation defects of haematopoiesis. *J Gen Virol* 1989; 70:1203-1215.

[17]Savard M, Belanger C, Tardif M, Gourde P, Flamand L, Gosselin J. Infection of primary human monocytes by Epstein-Barr virus. *J Virol* 2000; 74: 2612-2619.

[18]Shimakage M, Kimura M, Yanoma S, Ibe M, Yokota S, Tsujino G, Kozuka T, Dezawa T, Tamura S, Ohshima A, Yutsudo M, hakura A. Expression of latent and replicative-infection genes of Epstein-Barr virus in macrophage. *Arch Virol* 1999; 144: 157-166.

[19]Shimakage M, Sasagawa T, Kimura M, Shimakage T, Seto S, Kodama K, Sakamoto H. Expression of Epstein-Barr virus in Langerhans' cell histiocytosis. *Hum Pathol* 2004; 35: 862-868.

[20]Shimakage M. Langerhans cell histiocytosis and relationship with Epstein-Barr virusreply. Hum Pathol 2006; 37: 1509-1511.

[21]Lewis CE; Pollard JW. Distinct role of macrophages in different tumor microenvironments. *Cancer Res* 2006; 66: 605-612.

[22]Balkwill F, Mantovani A. Inflammation and cancer: back to Virchow? *Lancet* 2001; 357: 539-545.

[23]Coussens LM, Werb Z. Inflammation and cancer. *Nature* 2002; 420: 860-867.

[24]Aggarwal BB, Shishodia S, Sandur SK, Pandy MK, Sethi G. Inflammation and cancer: How hot is the link? *Biochem Phrmacol* 2006; 72: 1605-1612.

[25]Grivel J-C, Ito Y, Faga G, Santoro F, Shaheen F, Malnati MS, Fitzgerald W, Lusso P, Margolis L. Suppression of CCR5- but not CXCR4-tropic HIV-1 in lymphoid tissue by human herpesvirus 6. *Nature Med* 2001; 7: 1232-1235.

[26]Lisco A, Grivel JC, Biancotto A, Vanpouille C, Origgi F, Malnati MS, Schols D, Lusso P, Margolis LB. Viral interaction in human lymphoid tissue: Human herpesvirus 7 suppresses the replication of CCR5-tropic human immunodeficiency virus type 1 via CD4 modulation. *J Virol* 2007; 81: 708-717.

[27]Barton ES, White DW, Cathelyn JS, Brett-McClellan KA, Engle M, Diamond MD, Miller VL, Virgin HW. Herpesvirus latency confers symbiotic protection from bacterial infection. *Nature* 2007; 447: 326-330.

[28]Sasagawa T. Human papillomavirus infection and cervical cancer. *Biomed Rev* 2003; 14: 75-93.

[29]Hammes LS, Tekmal RR, Naud P, Edelweiss MI, Kirma N, Valente PT, Syrjanen KJ, Cunha-Filho JS. *Gynecologic Oncology* 2007; 105: 157-165.

INDEX

A

B

C

D

E

F

G

H

I

J

K

L

M

N

O

P

Q

R

S

T

U

V

W

Y